T0297975

GROSS POLLUTANT TRAPS to ENHANCE WATER QUALITY in MALAYSIA

Datin Prof. Ir. Dr. Lariyah Mohd Sidek
Ir. Hidayah Basri
AP Dr. Mohd Ahmed Hafez
Dato' Ir. Mohd Azmi Ismail
AP Dr Rohayu Che Omar

To order additional copies of this book, contact
Toll Free 800 101 2657 (Singapore)
Toll Free 1 800 81 7340 (Malaysia)
www.partridgepublishing.com/singapore
orders.singapore@partridgepublishing.com

ISBN
ISBN: 978-1-5437-5370-7 (sc)
ISBN: 978-1-5437-5372-1 (hc)
ISBN: 978-1-5437-5371-4 (e)

Print information available on the last page.

10/30/2019

Contents

ACKNOWLEDGEMENT

We are immensely grateful to the Department of Irrigation and Drainage (DID) Malaysia with unique thank you to the Pejabat Lembangan Sg Klang (PLSK) and Universiti Tenaga Nasional (UNITEN) for their full support to this research. This study is also supported by Ministry of Higher Education under Fundamental Research Grant Scheme 20170106FRGS. We would also like to show our gratitude to ZHL Engineers Sdn Bhd for sharing their pearls of wisdom with us during this research.

CHAPTER I:

INTRODUCTION

1.1 GENERAL INFORMATION

An increasing population due to urbanization contributes to the generation of increased amounts of rubbish and waste material. The uncontrolled dumping of rubbish can exert a huge impact on aesthetics and river water quality, both of which affect the quality of life in urban environments and increased management costs. The rising awareness of these factors has led to the implementation of gross pollutant management strategies as a holistic approach to water quality improvement (Q. Zhang et al., 2014).

Due to rapid and uncontrolled development in Klang Valley, Kuala Lumpur generate 3,500 tonnes of domestic and industrial waste a day and the cost of public cleansing and waste management is a whopping RM325mil a year, according to statistic from Public Cleansing Management Corporation (SWCorp)(The Star, 2016). The Department of Irrigation and Drainage reported that daily cleaning of log booms along the 110km stretch of the Sungai Kelang basin as part of the River of Life (RoL) project DID collects and disposes of between 10 and 13 tonnes of rubbish accumulated at its 10 log booms and 396 gross pollutant trap (GPT) in a month (The Star, 2018). An average total of RM500,000 a month was spent to maintain and clean the GPT. These accumulated pollutants are not only aesthetically unattractive, but also demonstrate environmentally threatening and devastating effect to the natural equilibrium, and impede the hydraulic performance of the urban drainage system (Ghani A. et. al., 2011).

There are a large number of factors which could affect the number of gross pollutants in stormwater. Among the influential variables which may influence gross pollutant loads includes catchment characteristics (a type of land use, population density, and affluence), rainfall characteristics, management practice (such as street sweeping/vacuuming, availability of collection bins and regularity of cleaning, recycling programs) and education and awareness programs.

The rising awareness about the degradation of river water quality by gross pollutants has led to the implementation of gross pollutant management strategies as a holistic approach for water quality improvement (Q. Zhang et al., 2014). The integration of both structural and non-structural measures is important to ensure the effectiveness of gross pollutant management. Structural measures are constructed in-transit treatments which separate and contain pollutants. The introduction of gross pollutant traps before entering the receiving water such as a pond, wetland, and river. This structural measure is based on the concept of "control-at-source" to control stormwater quantity and quality (Madhani et al., 2014).

The Malaysian Government's initiative under the River of Life (ROL) project has installed at least 528 units of GPTs to trap gross pollutants at the end of the drainage system and to trap gross pollutants before entering river system. The installation of GPTs is described in Key Initiatives No. 4 for River Cleaning under the ROL project. An effective stormwater management program should consist of strategies that will achieve the goal at the minimum cost, thereby making the best use of limited available resources (Brown et al., 1999).

There are now numbers of trapping devices with different trapping mechanism available in the market. Efforts have been made by the authority to install the Gross Pollutant Trap (GPT) to trap gross pollutants from entering the river system. It has been proven by many studies that these devices can reduce a significant amount of gross pollutants before entering the river system. In Malaysia, Chapter 10 in Manual Saliran Mesra Alam (MSMA) 2nd Edition suggests the use of GPTs at the end of every drain to trap gross pollutants before entering main river system (DID, 2012).

The management of stormwater is an extensive and costly exercise, and most local councils and water authorities do not possess sufficient funds for adequate stormwater management, especially for maintenance purpose. Tools to optimize the placement of traps are currently not available. As is not feasible to install gross pollutant traps in all catchments within a drainage system, a tool to estimate the optimal number and placement of traps is proposed, taking into the identification of the optimal number of GPT to be installed in a catchment is in line with the objective of the River of Life (ROL) project.

However, it is also important to note that these devices only function well with periodic and proper maintenance method. The operation of gross pollutant devices is governed by a number of confounding factors, including catchment size, pollutant load, type of drainage system, and cost. However, the performance of GPTs is not fully tested on real practical applications under the influence of tropical climate where high rainfall intensity in short duration prevails. (L. Sidek, 2015). The performance of GPTs is strongly dependent upon the specific site criteria including type of land use, hydrological regime and maintenance frequency. As maintenance cost is significant in the life cycle cost of GPTs, the local authority is facing issues of conducting proper maintenance frequency for installed GPTs, resulting in system clogging which leads to flooding and contaminating the water at the downstream area.

In view of this challenge, the Humid Tropics Centre (HTC) has appointed UNITEN R&D Sdn Bhd (URND) to provide services to carry out the Research on Performance and Optimum Numbers of Gross Pollutant Trap (GPT) Trapping Devices for River of Life (ROL). In this study, URND is required to review and enhance the existing GPTs inventory system, planning & management tool for managing GPTs; to optimize the maintenance cost in effective ways without affecting the normal maintenance operation system and structure and to develop the training kits and module curricula on the standard of all the Gross Pollutant Trap based on their types and components in line with MSMA Edition 2.

1.2 PROBLEM STATEMENT

Rapid urbanization has known to have several adverse impacts towards the hydrological cycle due to increased impervious surface and degradation of water quality in stormwater runoff. Waterways located in or near the urban area are adversely influenced by solids in stormwater runoff (Roesner, L.A., 2007).

Gross pollutants are defined as debris items larger than 5 mm (Allison et al. 1997a). Gross pollutants are transported into the water conveyance system during a storm event. Accumulation deposits of gross pollutants degrade the receiving water quality, blockage of conveyance system that leads to flooding and reduce the waterway aesthetic value.

Gross Pollutant Traps (GPT) is designed to capture and retain gross pollutants from entering a river system (Jehangir T. Madhani, 2014). Installation of GPT is a structural way to reduce gross pollutants in water conveyance system. Usually, GPT is installed at the end of the drainage system before entering the river system. However, it is important to understand that tropical climate always related to the situation where high rainfall intensity occurs in short duration. The accumulation of pollutant without proper maintenance activities result in clogging of waterways, which contributes to the flooding problem. Another problem that arises is the frequent maintenance requirement, which affects the life cycle cost of a GPT.

Chapter 10 in MSMA Manual 2nd Edition suggested the installation of GPT as one of water quality treatment practice. However, it was found that limited information is available on the GPT design criteria, description, efficiencies as well as operation and maintenance of available GPT based on local data and experience. Most of the data are from Australia, which is based on Australian local data. This study aims to conduct an assessment of GPT installed in an urban area in Malaysia. Data collection collected from study area able to provide local data such as the relationship of rainfall with gross pollutant load, gross pollutant load from a different type of land use, the relationship between population and pollutant load and estimation of life cycle cost.

Another issue on GPTs is the limited data sharing to the public in terms of its performance. It is essential to know the performance of GPTs to ensure the trap is suitable to be installed under the Malaysian condition. Thus, this study aims to suggest an efficient method in the management of gross pollutants. The local data collected finally become input to Innovative Gross Pollutant Management Strategies Knowledge Database, which helps engineers or local authorities to manage the GPTs efficiently. Furthermore, the knowledge management portal ensures knowledge sharing and the establishment of a knowledge bank for Gross Pollutant under Malaysian condition.

1.3 PROJECT OBJECTIVES

The study is about the effectiveness and performance of 375 unit of proprietary Gross Pollutant Traps (GPTs) stormwater quality control in the urban areas. The study is been conducted by providing a management and planning tool for effective management of the gross pollutants in the urban areas under the sub-catchment in River of Life (ROL) project, namely Sungai Klang, and Sungai Sering. The objectives of this research are:

1. To review and enhance the existing GPTs inventory system, planning & management tool for managing GPTs;
2. To optimize the maintenance cost in effective ways without affecting the normal maintenance operation system and structure;
3. To develop the training kits and module curricula on the standard of all the Gross Pollutant Trap based on their types and components in line with MSMA Edition 2.

1.4 STUDY AREA

The Klang River Basin is the most densely populated region in the country. The basin encompasses the Federal Territory of Kuala Lumpur and parts of Selangor districts of Hulu Langat, Gombak, Petaling, Sepang, Klang and Kuala Langat. The local councils that are within the basin are Dewan Bandaraya Kuala Lumpur, Majlis Bandaraya Shah Alam, Majlis Bandaraya Petaling Jaya, Majlis Perbandaran Ampang Jaya, Majlis Perbandaran Selayang, Majlis Perbandaran Subang Jaya, Majlis Perbandaran Sepang, Majlis Perbandaran Klang and Majlis Daerah Kuala Langat.

The Klang River Basin has radically transformed from its previous natural forested condition due to the various human activities. These were carried out in pursuit of urban and industrial development, mining and agriculture through activities that include the clearing of forests, cultivation, construction of dams, river canalization works, drainage of swamps, construction of roads, railways and other infrastructure and agricultural projects. Development in the Klang River Basin had concentrated on the rivers banks and their floodplains to gain ready access to water for domestic, industrial and agricultural purposes, river transport, fertile floodplain soils and for the convenient disposal of effluents and refuse. As a result of rapid land development, these areas have become prone to flooding.

1.4.1 DESCRIPTION OF RIVER BASIN

The Klang River Basin is located at the middle of the west coast of Peninsular Malaysia. It fully encompasses the Federal Territory of Kuala Lumpur (WPKL), and parts the state of Selangor for a total area of approximately 1,288km^2. It discharges into the Straits of Malacca, as shown in Figure 1.1.

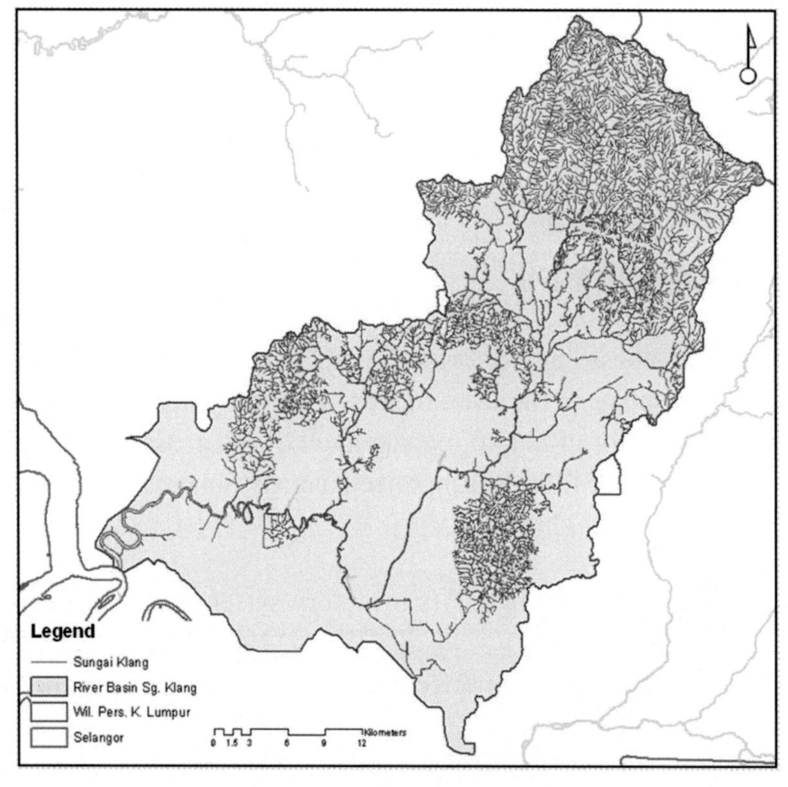

Figure 1.1 Klang River Basin

The main tributary system at the upper reach of the Klang River Basin includes that of Batu, Gombak and Upper Klang as illustrated in Figure 1.2. Except for Gombak, the upper reaches of the other two rivers are dammed for flood mitigation and water supply purposes.

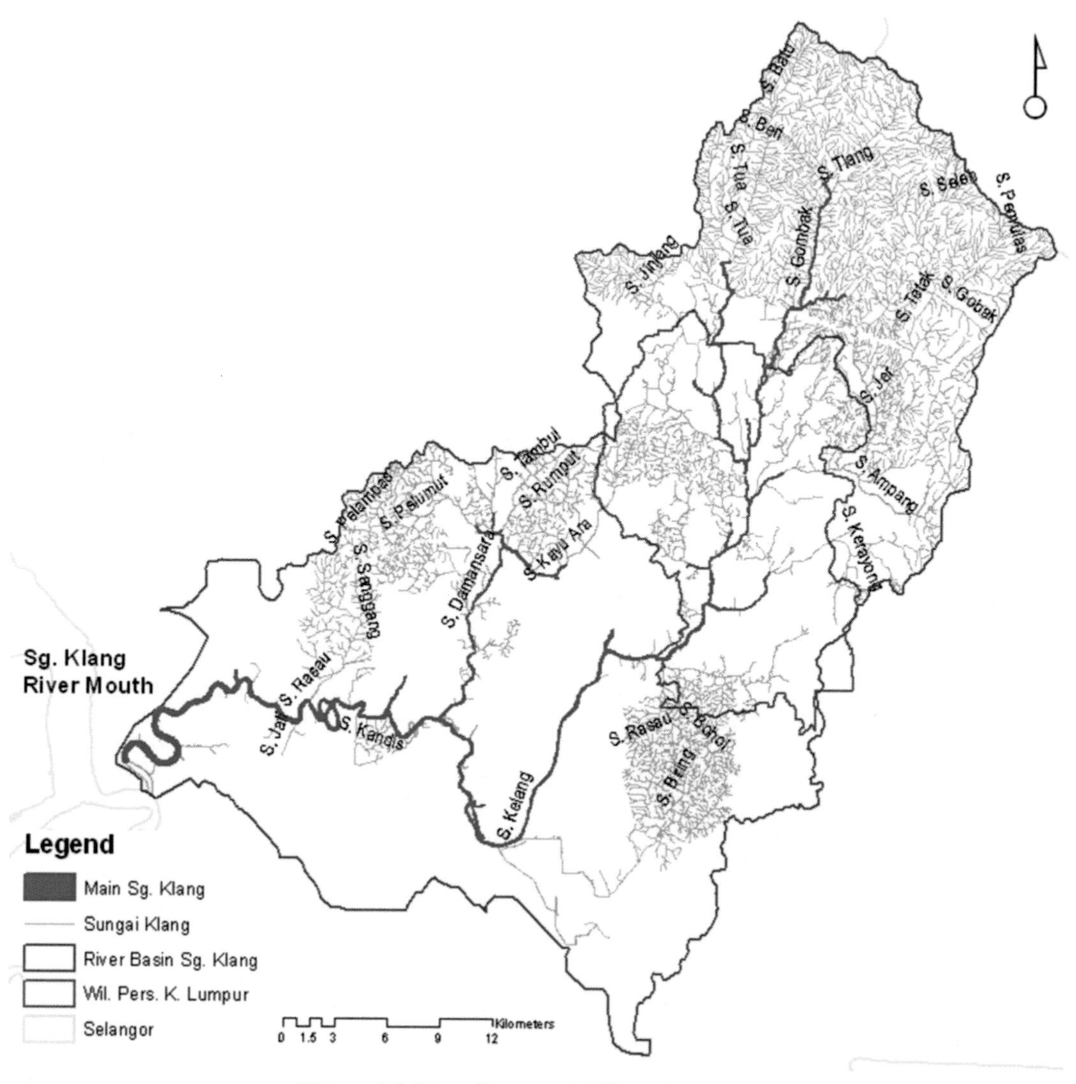

Figure 1.2 River Systems in Klang River Basin

1.4.2 RIVER OF LIFE (ROL) BOUNDARY

ROL project is a Malaysian Government idea under the Economic Transformation Program to reach greater Klang Valley by transforming the Klang River into a vibrant and livable waterfront by the year 2020. The proposed study area covers the whole ROL boundary (approximately 560 km^2) is located in Sungai Klang upper catchment, which flows through Kuala Lumpur and Selangor in Malaysia and eventually flows into the Straits of Malacca as shown in Figure 1.3. This study focuses on the whole catchments in ROL Boundary which include Sungai Klang, Sungai Kemensah, Sungai Gisir, Sungai Sering, Sungai Batu, Sungai Jinjang, Sungai Kerayong, Sungai Keroh, Sungai Bunus, Sungai Ampang, and Sungai Gombak.

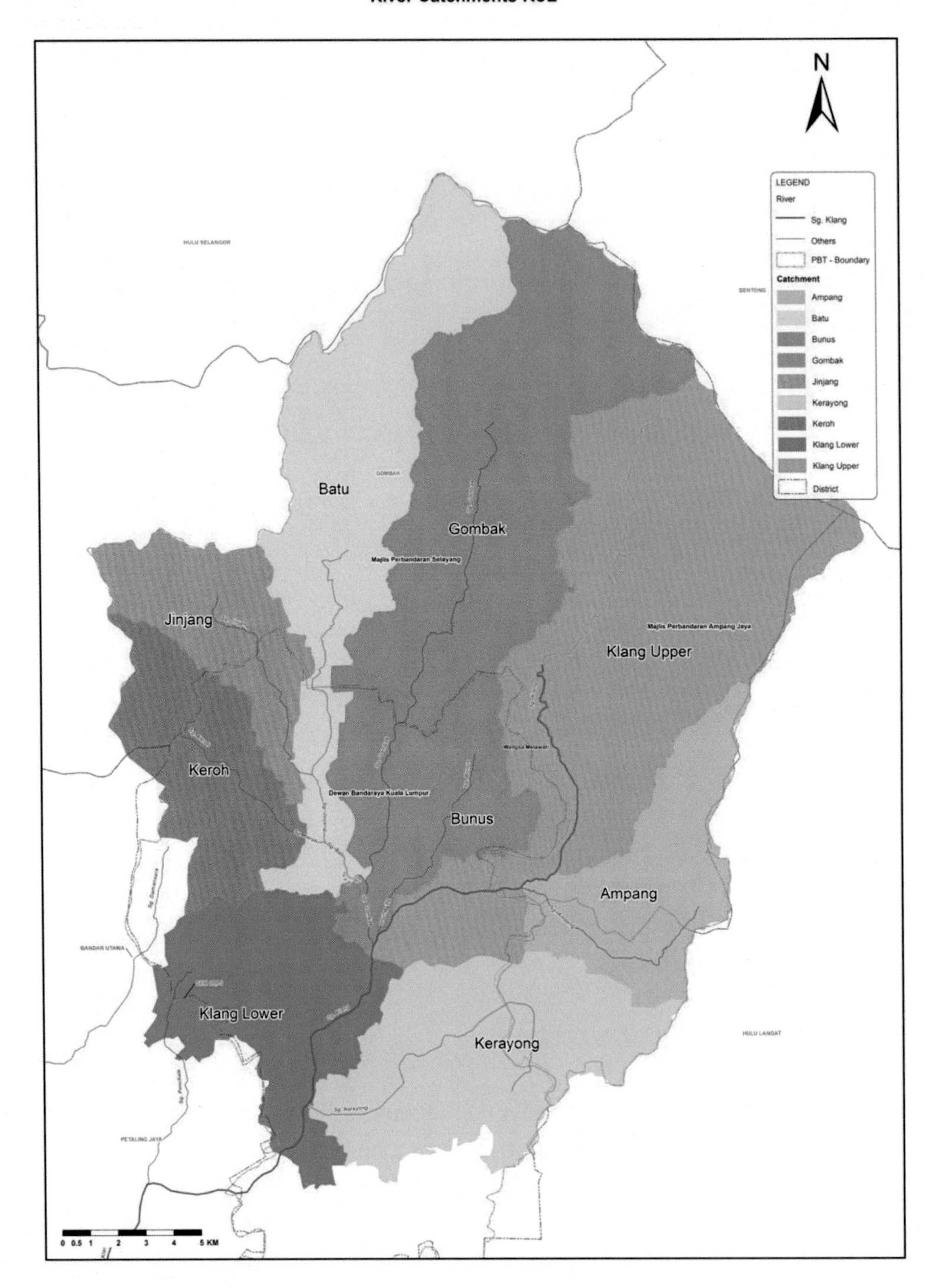

Figure 1.3 The study location within the ROL boundary

1.4.3 GPT LOCATION IN ROL CATCHMENT

To date, there is 401 number of GPTs have been installed within the ROL boundary. However, there are only 375 units of proprietary GPTs operated and maintained by PLSK and 6 units operated by DBKL. Some of the GPTs were not maintained anymore due to damaged and buried under the construction site. Figure 1.4 shows the location of proprietary GPTs for ROL catchment. The total number of proprietary GPTs installed in each catchment is tabulated in Table 1.1.

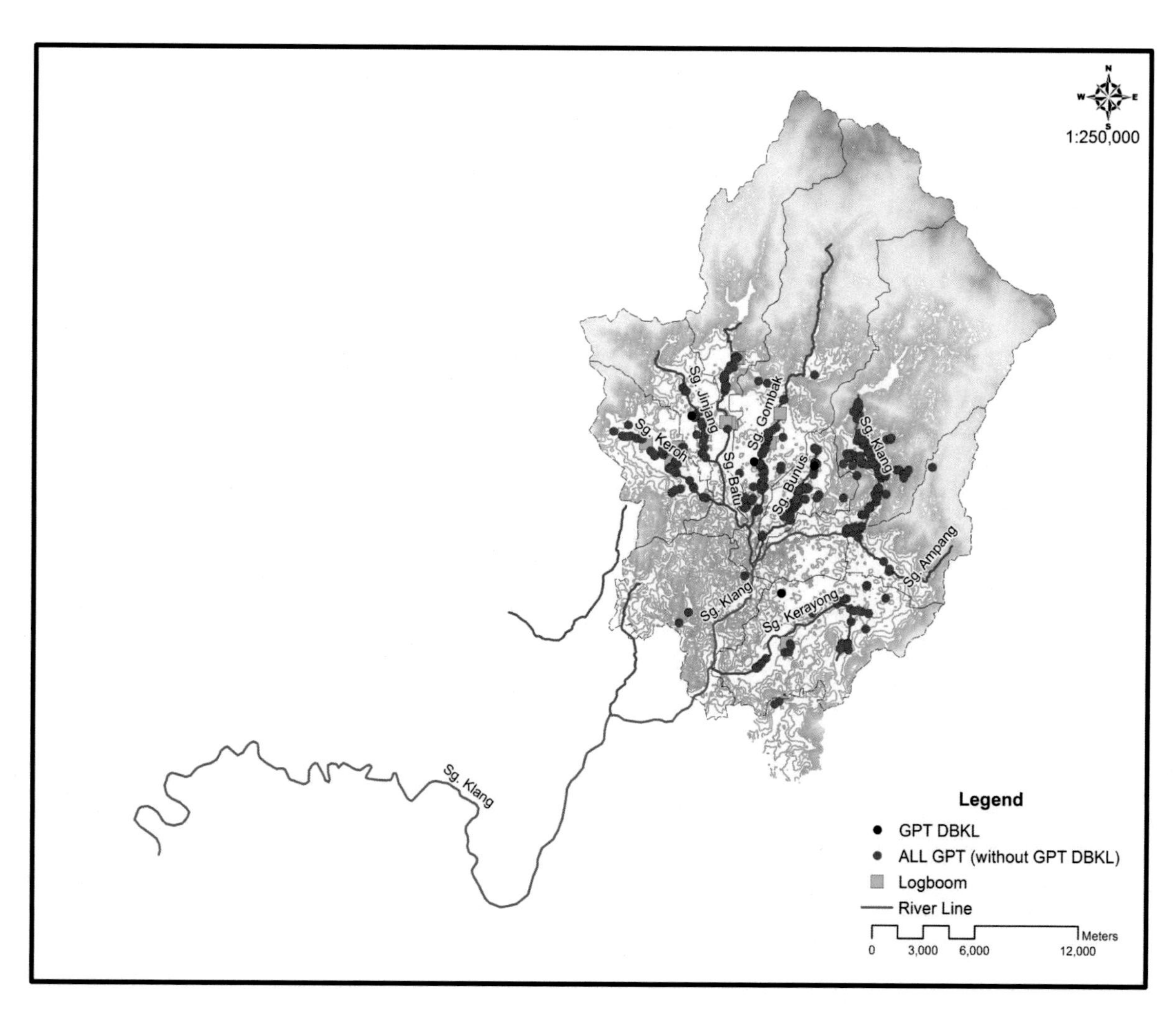

Figure 1.4 GPTs Location within ROL Catchment

Table 1.1 Total Numbers of GPTs in ROL

Installed By	Location/Package	Total
JPS	**Sg Klang**	88
	Sg Gisir	31
	Sg Sering	19
	Sg Kemensah	10
	Sg Kerayong	38
	Sg Gombak	64
	Sg Batu	20
	Sg Bunus	55
	Sg Keroh	47
	Sg Ampang	5
	Sg Jinjang	18
DBKL	**DBKL**	6
	Total Installed	**JPS: 395 DBKL: 6**
	Total Damaged GPT	**JPS: 20 DBKL: -**
	Total Maintained	**JPS: 375 DBKL: 6**

There are 8 types of proprietary GPTs installed in River of Life Project which are Continuous Deflective Separation (CDS), Downstream Defender (DD), Solid Interceptor (SI), CleansAll (CA), Neutralizing Turbulent Vortex System (NTVS), HumeGard (HUME), Ecosol (ECO) and PVT. Table 1.2 shows the model specification and the approximation dimension for 8 types of proprietary GPTs. This information is essential to determine the maintenance frequency for each type of GPTs.

Table 1.2 Model Specifications of GPTs

Type	Introduction	Image
Continuous Deflective Separation (CDS)	– CDS technology is ideally suited to the removal of physical pollutants from stormwater targeting the removal of gross solids and sediment, Effective over a wide range of treatment flows, No clogging / Blinding, Low life cycle costs. – Nonblinding/clogging screen technology High rate physical separation, screen and vortex scparation Guaranteed treatment flow rate, Separate treatment and storage zones, Pollutants stored out of flow path, reduced re-suspension Benefits of Direct Screening and Vortex Separation.	

Specification				
CDS Model No	**Catchment Area (ha)**	**Pollution Storage (m^3)**	**Underground Footprint (m)**	**Ground Level footprint Diameter (m)**
F0506	≤ 1	0.50	1.0x1.0	0.9
P0708	≤ 2	1.80	1.5x1.5	1.3
F0908	≤ 4	2.65	1.5x1.5	1.3
F0912	≤ 6	2.95	1.5x1.5	1.3
P1009	≤ 8	3.35	2.0x2.0	1.3
P1012	≤ 12	3.60	2.0x2.0	1.3
P1015	≤ 18	3.80	2.0x2.0	1.3
P1512	≤ 20	7.10	2.6x2.6	1.8
P2018	≤ 45	19.00	3.5x3.5	2.3
P2028	≤ 75	21.50	3.5x3.5	2.3
P3018	≤ 100	28.50	6.5x6.5	3.5
P3024	≤ 150	32.75	6.5x6.5	3.5
P3030	≤ 200	37.00	6.5x6.5	3.5
C4527	200-250	58.50	8.0x8.0	5.0
TWIN	≤ 500	Variable	Variable	Variable

Type	Introduction	Image
Downstream Defender (DD)	– Downstream Defender is an advanced vortex separator used to treat stormwater runoff in pretreatment or stand-alone applications. Its unique flow-modifying internal components distinguish the Downstream Defender® from conventional and simple swirl separators that typically bypass untreated peak flows to prevent washout of captured pollutants. Its wide treatment flow range, low head loss, small footprint and low profile make it a compact and economical solution for capturing nonpoint source pollution.	

Specification						
Model No and Diameter	**Peak Treatment Flow Rate**	**Max Pipe Diameter**	**Oil Storage Capacity**	**Sediment Storage Capacity**	**Min Distance from Outlet Invert to Top of Rim**	**Standard Height from Outlet Invert to Sump Floor**
ft	**L/s**	**mm**	**L**	**m^3**	**m**	**m**
4	85	300	265	0.53	0.85	1.25
6	227	450	818	1.61	0.98	1.80
8	425	600	2044	3.56	1.28	2.35
10	708	750	3975	6.65	1.52	2.85
12	1076	900	6700	11.24	1.71	3.41

Type	Introduction	Image
Solid Interceptor (SI)	– Solid Interceptor protects downstream waterways by targeting and removing the trash, debris, and coarse contaminants at high flow rates. – Solid Interceptor is ideally suited for installation at the collection source in small catchment areas and is designed to remove gross pollutants, organic waste, silt, sediment and oils. The polymer construction makes it light and easy to use.	

Specification					
Size	**Max Discharge Capacity (m^3/s)**	**Max Discharge Capacity (I/S)**	**Min. Sit Trapping Capacity (m^3)**	**Max Floatable Trash Trapping Capacity (m^3)**	**Max. Oil Trapping capacity (L)**
1.5	0.112	112	18kg	0.5	Up to 12
1.8	0.178	178	22kg	0.7	Up to 14
2.1	0.261	261	25kg	0.9	Up to 17
2.4	0.402	402	29kg	1.2	Up to 19

Type	Introduction	Image
CleansAll (CA)	– CleansAll units are ideal in the defense against a wide range of pollutants entering stormwater networks. – Designed to treat large flows in-line or end-of-line, protects the environment by removing gross pollutants, and suspended solids before they can reach the ocean, waterways, or wetlands. – CleansAll units will capture 100% of all gross pollutants at 3mm and above, or below the designed treatment flow rate. On average, this represents greater than 95% of gross pollutants during all storm events. Sizes are available to suit most applications.	

Specification

Unit Type	Unit Name	Treatment	Inlet/ Outlet	Min Depth	X	Y	Z	Storage (m^3)		
		Flow	Pipe Range	To invert	Length	Width	(min)		Sediment	Oil
		(L/s)	(mm)	(mm)	(mm)	(mm)	(mm)	Basket(s)	Sump	Storage
CleansAll 375	CL375	90	300-600	990	2000	1690	2570	0.20	0.26	0.26
CleansAll 600	CL600	320	450-750	1070	2730	2365	3170	1.10	1.09	1.09
CleansAll 750	CL750	750	750-1200	1510	4000	3310	3610	1.80	3.16	3.16
CleansAll 900	CL900	928	750-1200	1510	4000	3310	4370	3.30	3.16	3.16
CleansAll 1200	CL1200	2200	1200-1650	2210	6010	5175	5090	5.70	10.95	10.95
CleansAll 1350	CL1350	2732	1200-1650	2210	6010	5175	5690	9.20	10.95	10.95

Type	Introduction	Image
Neutralizing Turbulent Vortex System (NTVS)	– Importantly and unlike other systems, the NTVS continues to treat the specified TFR even under high flow conditions. While the composition and quantity of pollutants vary depending on the catchment size, area, land use, and rainfall patterns, the NTVS typically captures the following pollutants.	

Specification					
Model No.	**Footprint Size (m x m)**	**Depth from Surface (m)**	**Treatable Flow Rate (m^3/s)**	**Equivalent Catchment Area (ha)**	**Holding Capacity (m^3)**
NTVS 1000	1300 x 2300	2350	0.03 – 0.04	0.19	1
NTVS 1500	1900 x 2650	2450	0.10 – 0.11	0.54	3
NTVS 2000	2450 x 3500	2600	0.20 – 0.22	1.1	5
NTVS 3000	3500 x 5000	3600	0.58 – 0.61	3.0	13
NTVS 4000	4600 x 6600	4650	1.22 – 1.26	6.3	29
NTVS 5000	5700 x 8200	5450	2.14 – 2.19	11.0	53
NTVS 6000	6800 x 9800	6100	3.39 – 3.46	17.5	86

Type	Introduction	Image
HumeGard (Hume)	– The Humegard GPT is a pollution control device that is specifically designed to remove gross pollutants and coarse sediments (≥ 150 microns), from stormwater runoff. – This system is designed for residential and commercial developments where litter and sediment are the predominant pollutants. – The floating boom and bypass chamber enable the continued capture of floating material, even during peak flows, and the unit also prevents the re-suspension and release of trapped materials during subsequent storm events.	

Specification					
HumeGard	**Pipe diameter of box culvert width (mm)**	**Width (mm)**	**Length (m)**	**Depth from pipe invert (m)**	**Total pollutant capacity (m3)**
HG12	300-375	1.8	2.0	1.39	2.0
HG15	450-525	1.8	2.0	1.36	2.0
HG18	525-675	2.0	2.1	1.10	2.5
HG22	675-825	2.5	2.5	1.25	6.0
HG24	600-825	2.7	2.5	1.70	6.8
HG27	750	3.0	2.5	1.60	6.7
HG30	750-900	3.4	2.5	2.20	10.6
HG30A	900	3.4	2.5	1.90	9.7
HG35	900	3.9	2.5	2.00	11.3
HG35A	1050	3.9	2.5	1.80	10.2
HG40	900	4.4	2.9	2.00	14.5
HG40A	1050	4.4	2.9	1.80	13.2
HG40B	1200	4.4	2.9	1.60	8.9
HG45	1200	4.9	2.9	2.10	18.3
HG45A	1350	4.9	3.2	1.90	18.0
HG50 and larger	>1500	5.3	3.5	2.00	>20.0

Type	Introduction	Image
Ecosol (ECO)	– Ecosol GPT in-line/end-of-line primary treatment solution is a non-blocking tangential screen filtration system that captures and retains 99% of pollutants larger than 211um and 90% of particles larger the 152um at its design TFR. – There are two types low flow Ecosol in line GPT for primary and secondary treatment of stormwater at low flow velocities and, the high flow Ecosol in line GPT for primary treatment of stormwater at high flow velocities.	

Specification					
Ecosol Product Code	**Inlet/Outlet Pipe Diameter**	**Approx. External Dimensions (L*W*D from invert)**	**Holding Capacities**		
			Solid pollutants	**Free Oil and Grease**	**Water**
		mm	**(m^3)**	**(liters)**	**(liters)**
GPT 4200	Up to 375 mm	2200x900x750	0.23	268	667
GPT 4300	150 to 600mm	2700x1350x750	0.32	469	1181
GPT 4450	225 to 900mm	3600x1650x1050	1.03	1347	3348
GPT 4600	300 to 1200mm	4500x1950x1350	2.43	2994	7211
GPT 4750	450 to 1350mm	5600x2300x1650	4.83	5711	13608
GPT 4900	600 to 1650mm	6500x2600x1975	8.30	9576	22768
GPT 41050	750 to 1800mm	7450x2950x2300	13.11	14850	35262
GPT 41200	900 to 2100mm	8630x3300x2625	19.52	22793	51698
GPT 41350	1050 to 2400mm	9700x3700x2950	27.70	30578	72495
GPT 41500	1200 to 2400mm	10680x4000x3250	37.94	41491	98317
GPT 41800	1350to 2400mm	12730x4700x3900	65.33	70452	166836

CHAPTER II:

GROSS POLLUTANT MANAGEMENT STRATEGIES

2.1 INTRODUCTION

Rapid urbanization in Malaysia with the construction of new urban conglomeration tends to change the hydrologic, hydraulic, and environmental characteristics of catchments. Apart from the physical impacts of flooding, urbanization also resulted in water quality degradation of urban river and other receiving waters. The quality of runoff is influenced by many factors such as the type of land use, waste disposal and sanitation practices. In Malaysia, gross pollutants such as litter, debris and sediments are one of the main causes of river pollution and flooding problem. Despite education, awareness and street cleaning programs, a large number of gross pollutants (litter and debris greater than 5mm in size) are reaching and degrading the rivers. As an approach to improve river water quality, MSMA 2nd Edition has introduced the use of GPTs to trap gross pollutant in urban waterways. There are many types of proprietary GPTs available in the market. Each GPT has its unique performance depending on several factors, which includes catchment characteristics and cost. This chapter provides a literature review on various key topics related to this study, which covers gross pollutant management strategies.

2.2 GROSS POLLUTANT

The term "gross pollutant", when used in connection with stormwater drainage systems, include litter, debris and coarse sediments (Fitzgerald, 2010). Litter is defined as human-derived material including paper, plastics, metals, glass and cloth. Debris is defined as any organic material transported by stormwater (such as leaves, twigs and grass clippings). Sediments are defined as inorganic particulates. While all gross pollutants are not 100% human-derived, human activities are likely responsible for an exponential increase in pollutants over predevelopment conditions. Therefore, it is important to understand the factors causing accumulation of gross pollutants to manage them effectively, especially in urban waterways (Sidek et al., 2011).

2.2.1 *TYPE OF PROPRIETARY GPT'S*

Table 2.1 Proprietary Gross Pollutant Traps

Gross Pollutant Treatment Devices	**Description**	**Catchment Size**	**Treatable Flow**	**Efficiency** (pollutant capture)	**Capital Cost** (Supply and Install)	**Maintenance Cost** (per service)	Country
Removal Mechanism: Litter Basket							
Litter Basket	<u>Description</u> – Design with flatter grades where the basket sits on the pit base just below the inlet pipe invert. – Made of a wire mesh or plastic 'basket' installed in a stormwater pit to collect litter from a paved surface (litter basket) – This design also incorporates a vertical trash grate within the pit to minimize blocking by gross pollutants – The installation of the second length of pipeline parallel and slightly higher than the main line. This second line acts to convey flow bypass that results once the basket blocks.	1 - 2 ha	< 1000 L/s	Litter – (30-50%) Sediment: Coarse – (30-50%) Medium – (10-30%) Fine – none Oil and Grease – (30-80%) if oil sock or pillow incorporated into the device	AUD 50,000 – AUD 130,000	AUD 1,200 per year by Vacuum Truck Monthly	Australia

Gross Pollutant Treatment Devices	**Description**	**Catchment Size**	**Treatable Flow**	**Efficiency** (pollutant capture)	**Capital Cost** (Supply and Install)	**Maintenance Cost** (per service)	Country
Litter Pit	Description – Consists of a removable pit basket which is positioned on an elevated framework within either an existing or newly constructed pit. – The basket is positioned below the invert of the inlet pipe, and it is elevated enough to allow flow to pass under the basket at the outlet. – Stormwater entering the pitfalls under gravity through an angled trash grate and out beneath the basket. – The hydraulic push of the stormwater across the grate allows suspended and floatable debris to be caught in the basket below. This design allows for minimal blocking and flows obstruction in the pit.	150 ha	< 3000 L/s	Litter – (30-50%) Sediment: Coarse – (30-50%) Medium – (10-30%) Fine – none Oil and Grease – (30-80%) if oil sock or pillow incorporated into the device	AUD 10,000- AUD 50,000	AUD 4,000 per year by Vacuum Truck or Hand per month	Australia

Gross Pollutant Treatment Devices	Description	Catchment Size	Treatable Flow	Efficiency (pollutant capture)	Capital Cost (Supply and Install)	Maintenance Cost (per service)	Country
Ski Jump	Description – To remove litter and debris from stormwater. – It is installed to the outfall of the existing pipe system with drop 300 to 400 mm. – The contained shallower basket that captured the materials from the stormwater.	< 50 ha	< 3000 L/s	Litter – (80-100%) Sediment: Course – (40-70%) Medium – (10-30%) Fine – none Nutrients – none Metals – none Oil & Grease – none	AUD 2,800 to AUD 17,000	2 – 3 months by Hand or Vacuum Truck AUD 6,000 – AUD 10,000 per year	Australia
Side Entry Pit Traps (SEPTs)	Description – placed in the entrance to drains from road gutters – consists of a mild steel basket mesh measuring between 5-20 mm. – The basket has an open-top which receives water and pollutants with the latter trapped into it. – The traps are stored in a space at the rear of the pit to provide a flow path for high flows.	≤ 1 ha	nil	Moderate (depend on the number of devices used)	AUD1,000-AUD5,000	AUD300 – AUD400 per service RM 200 – RM 300	Malaysia Australia

Gross Pollutant Treatment Devices	Description	Catchment Size	Treatable Flow	Efficiency (pollutant capture)	Capital Cost (Supply and Install)	Maintenance Cost (per service)	Country
Rocla Cleans all Traps	– trap gross pollutants, oils and sediment – divert treatable flow into a circular chamber – A sediment sump is located downstream the circular chamber – An internal bypass system enables peak storm events to bypass the collection chamber without remobilizing the captured pollutants	≤ 200 ha	90 L/s to 6000 L/s	Litter – (80-100%) Sediment: Coarse – (80-100%) Medium – (30-50%) Fine – (10-30%) Oil and Grease – (≤10%) Bound to captured sediment Nutrient – (≤10%) Bound to captured sediment	AUD30,000-AUD200,00	AUD 1,500 –AUD 2,500 per service	Australia
Ecosol Traps i. At-source RSF 100 ii. At-source RSF GSP iii. End-of-line RSF 1000	Description – Trap litter at entry points, end-of-line, inline of the drainage system – consists of a capture basket – consist of an overflow by-pass flap(s) – As the basket is almost full of stormwater, the by-pass flap(s) begins to open in response to the incoming flow. – This is to prevent the possibility of flood occur. – When the flow ceases, the flap(s) will close	≤ 200 ha	≤ 4000 L/s	Very High	AUD30,000-AUD250,000	AUD 2,000 – AUD 3,400 per service by vacuum truck	Australia

Gross Pollutant Treatment Devices	**Description**	**Catchment Size**	**Treatable Flow**	**Efficiency** (pollutant capture)	**Capital Cost** (Supply and Install)	**Maintenance Cost** (per service)	**Country**
iv. In-line /end of line RSF 4000	<u>In-line/end-of-line RSF 4000</u> – This unit removes and retains solid pollutants, free oils, grease, and fine sediments from stormwater flows. – It can be fitted to conduits of almost any size or shape and can also accept flows from open channels. – Stormwater and pollutants are diverted into a litter-collection basket by a hydraulically driven barrier created by the water exiting the basket. – Pollutants are retained in the basket by direct filtration.	10–200 ha	≤ 4000 L/s	Litter – (80-100%) Sediment-(80- 100%) Oil & Grease –(80 – 100%) Retain more than 270 tonnes of pollutants and up to 212,000 liters of free oils and grease conveyed by stormwater flows each rain event	AUD200,000 – AUD 250,000	3 – 5 months per cleaning by vacuum truck	Australia, Putrajaya, Malaysia.

Gross Pollutant Treatment Devices	Description	Catchment Size	Treatable Flow	Efficiency (pollutant capture)	Capital Cost (Supply and Install)	Maintenance Cost (per service)	Country
Removal Mechanism: Hydrodynamic Separators							
Continuous Deflective Separator (CDS) i. (inline unit) ii. end-off-line units	<u>Description</u> – Trap litter and coarse sediments – Utilized the vortex action which pulls pollutants toward the center of a collection chamber – Solids within the separation chamber are kept in continuous motion preventing from 'blocking' the screen. – Water passes through the screen and flows downstream	≤ 200 ha	≤ 4000 L/s	Very High	AUD30,000-AUD300,000	AUD1,000 - AUD 2,000 per service RM 3,000 – RM 3,500 per service	Australia Malaysia

Gross Pollutant Treatment Devices	Description	Catchment Size	Treatable Flow	Efficiency (pollutant capture)	Capital Cost (Supply and Install)	Maintenance Cost (per service)	Country
Rocla Cleansall Traps	Description – trap gross pollutants, oils and sediment – divert treatable flow into a circular chamber – A sediment sump is located downstream the circular chamber – An internal bypass system enables peak storm events to bypass the collection chamber without remobilizing the captured pollutants	≤ 200 ha	90 L/s to 6000 L/s	Litter – (80-100%) Sediment: Coarse – (80-100%) Medium – (30-50%) Fine – (10-30%) Oil and Grease – (≤10%) Bound to captured sediment Nutrient – (≤10%) Bound to captured sediment	AUD 30,000 - AUD 200,00	AUD 1,500 – AUD 2,500	Australia
Baysaver	Description – The Baysaver system is essentially untested at this time. – One study has been conducted on the hydraulics of the system, however, no litter studies have been performed. – Baysaver units hold a permanent pool of water. – This has the potential for mosquitoes to breed and therefore requires additional inspection be the vector control agencies to ensure there is no mosquito breeding or to provide abatement.	nil	nil	nil	AUD 7,000 - AUD 10,000 for the smallest model AUD 13,000 - AUD 20,000 for largest model	Cleaned by vacuum truck The maintenance cost depends on the size of the unit	Australia

Gross Pollutant Treatment Devices	Description	Catchment Size	Treatable Flow	Efficiency (pollutant capture)	Capital Cost (Supply and Install)	Maintenance Cost (per service)	Country
Vortechs	Description – Trap sediment, debris, oil and grease and other pollutants. – Stormwater flows enter the unit tangentially to the grit chamber, which allows a gentle swirling motion. – As polluted water circles within the grit chamber, pollutants migrate toward the center of the unit where velocities are the lowest. – The majority of settable solids are left behind as stormwater exits the grit chamber through two apertures on the perimeter of the chamber. – There is a 5 mm mesh screen sloped between the grit chamber and the oil baffle wall to separate the trash and debris. – Next, buoyant debris and oil and grease are separated from water flowing under the baffle wall due to their relatively low specific gravity. – As stormwater exits the system through the flow control wall and finally through the outlet pipe, – Appropriate in residential subdivisions, commercial areas and in retrofits to existing stormwater drainage systems.	nil	nil	No studies have been done on the trash and debris removal efficiency.	AUD 10,500 -AUD $40,000	Give limited field performance data nil	Australia

Gross Pollutant Treatment Devices	Description	Catchment Size	Treatable Flow	Efficiency (pollutant capture)	Capital Cost (Supply and Install)	Maintenance Cost (per service)	Country
Trash Rack / Screen Devices							
Inclined GSRD 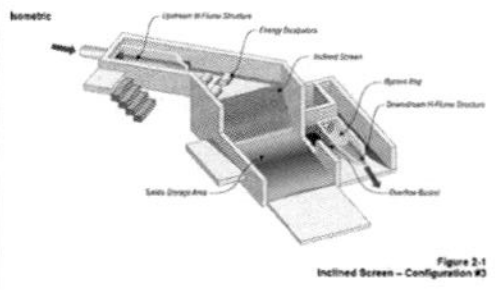	<u>Description</u> – this device uses a wedge-wire screen to remove gross solids. – The GSRD is configured with an upstream shelf to slow down the flow and evenly distribute it over the length of the screen. – The flow overtops a weir and falls through an inclined screen located after the upstream shelf. – After passing through the screen, the flow exits the GSRD. Gross solids are retained in a confined storage area that can be accessed by maintenance equipment.	< 150 ha	30 m^3 / s	Litter – (80-100%) Sediment: Course – (40-70%) Medium – (10-30%) Fine – none Nutrients – none Metals – 15 % Oil & Grease – Bound to captured litters	AUD 110,000	AUD 3,000-AUD10,000 per year Every 1-3 months	Australia

Gross Pollutant Treatment Devices	Description	Catchment Size	Treatable Flow	Efficiency (pollutant capture)	Capital Cost (Supply and Install)	Maintenance Cost (per service)	Country
Baramy Traps	Description – Installed at the end of pipe traps – Trap litter and debris are trapped by directing the outflow from a stormwater pipe down an inclined screen – The waterfalls through the screen and is discharged beneath the litter chamber into the drainage system	10 – 100 ha	24 m^3 / s	Litter – (≤ 70 %) Course – none Medium – none Fine – none Nutrients – none Metals – none Oil & Grease – none	AUD 10,000-AUD 150,000	AUD 3,000-AUD10,000 per year Every 1-3 months by Front End Loader or Hand	Australia
Baramy Vane Deflector (in-line) Dual Vane Trap Deflector Trap	– Description – Installed at in-line open channel – Baramy Vane Deflector Trap is designed to treat large volume flows at medium to high velocities as evidenced in open stormwater channels. – Consists of horizontal deflector units placed at varying heights to guide pollutants toward the outside of the channel & into a holding bay located either inside or beside the channel. – Vertical vanes may also be used to supplement and increase the capture efficiency. – Water pressure will keep the captured pollutants contained and will drain dry upon abatement of the flow. – Deflector traps require no headloss and make use of any centrifugal forces acting on the water flow to assist the deflectors in separating the pollutants from the flow	< 200 ha	Up to 65 m^3 / s	Litter – (80-100%) Sediment: Course – (20 -30%) Medium – none Fine – none Nutrients – none Metals – (15 -30)% Oil & Grease – Bound to	AUD 3000 to AUD 23,000	AUD 1200 – AUD 2500	Australia

Gross Pollutant Treatment Devices	Description	Catchment Size	Treatable Flow	Efficiency (pollutant capture)	Capital Cost (Supply and Install)	Maintenance Cost (per service)	Country
Sediment Traps							
Sand Filters	<u>Description</u> – There are two components to be sized for large sand filters, namely the upstream settling (or pre-treatment) basin and the filter. – These components can be designed based on a design storm event, and high flows in excess of the design storm can be designed to bypass the filter.	< 2 ha	nil	Gross Pollutant – (should be performed prior to discharge of sand filter) Sediment: Course – (80 – 100%) Medium – (70 – 80%) Fine – (40 – 60%) Nutrients – (30 – 50%) Metals – (30 – 50%) Oil & Grease – (80–100%)	AUD 5,000 – AUD 50,000	Monthly (when filter material appears to be clogged) AUD 1,000 – AUD 5,000	Australia Malaysia

Gross Pollutant Treatment Devices	Description	Catchment Size	Treatable Flow	Efficiency (pollutant capture)	Capital Cost (Supply and Install)	Maintenance Cost (per service)	Country
Litter Booms / Bandalong Traps							
Bandalong Traps i. Bandalong Floating Litter Trap	<u>Description</u> – floating device installed along waterways to collect and retain floating litter, vegetation and other debris. – It is like a floating raft supported by polyethylene pipes allowing it to float – Bandalong is held in place by galvanized chains which are attached to ground anchors. – Their trapping efficiency enables direct floating litter to flow in through its one-way mesh gate where materials are captured and retained for removal. – In slow-moving waters accumulated pollutants captured by the trap become heavy and may cause the device to sink. Thus by using a system of counterweights prevent the device from sinking. – Whilst the paddles close the entry gate when water flow ceases or when there is a change in flow directing due to changes in tidal action and wind direction. <u>Limitations</u> – Can only collect floating gross pollutants – Visually obstructive – Block access to watercraft	≥ 250 ha	nil	Gross Pollutant – (10-40%) Sediment – none Nutrients – none Metal – none Oil & Grease - none	AUD 40,000 – AUD 100,000	Every 1-2 weeks by hand AUD 6,000-AUD 10,000 per year	Australia

ii. Bandalong Boom Systems	<u>Description</u> – The booms are typically installed across a waterway (pond, channel or creek) to collect floating and partly submerged litter and debris. – Used to remove floating litter. – Collect litters at a single location. – No hydraulic head loss. – The boom can rise and fall with changing the water level – The waterlogged and neutrally buoyant material, such as plastic bags are filtered and trapped under the boom by the flow velocity. – Consist of a hanging mesh skirt, weighted skirt (submerged beneath the hanging mesh skirt), and boom's collar. – These booms are attached to the bank of the waterway. <u>Limitation:</u> – Only traps floating litter and debris (maybe a small proportion of the total load) – Floating and neutrally buoyant litter can be swept under the skirt during high flows. – Impacts from large objects such as branches or boats can reduce boom effectiveness. – Litter can be blown over the boom's collar in high winds. – Maintenance can be difficult as most booms must be cleaned by boat. – The potential for vandalism. – The possibility of sinking due to marine growth. – Collected litter can move upstream along a tidal channel due to tidal flows. – Low visual amenity	≥ 100 ha	nil	Litter – (10-30%) Sediment – none Nutrients – none Metal – none Oil & Grease - none	AUD5,000 – AUD 10,000	Every 1-2 weeks by hand AUD 6,000- AUD 10,000 per year	Australia

Gross Pollutant Treatment Devices	Description	Catchment Size	Treatable Flow	Efficiency (pollutant capture)	Capital Cost (Supply and Install)	Maintenance Cost (per service)	Country
Drain Inlet Insert							
Bio-Clean Insert	– Water flows over the weir and into the removable basket, filtering trash and debris. – Hydrocarbon booms catch hydrocarbons entering the storm drain. The basket is located directly under the manhole. – For installation into a square catch basin, there is a left half and a right half that telescope together to adjust for size, which makes up the main body of the insert and mounts solidly to the catch basin wall with either drive pins. – The Curb Inlet Basket is made from the high-quality marine-grade fiberglass and stainless steel. It is designed to prevent floatables from escaping during heavy flows.	< 2 ha	nil	Litter – (10-30%) Sediment – none Nutrients – none Metal – none Oil & Grease - none	nil	AUD 200-AUD 500 Cleaned monthly by removing the manhole lid and vacuum or remove the basket	Australia
Drain Pac Insert	– The DrainPac™ is a flexible storm drain catchment and filtration liner designed to filter pollutants, debris, and solids prior to discharge into storm drain systems. – The filters must be cleaned, possibly after each rainfall, with a truck-mounted vacuum so that the debris does not clog storm drain. – The DrainPac™ adapts to any size or shape inlet. It may have limited roadway application because of clogging. – It targets heavy sediments, oil and grease.	< 2 ha	nil	Litter – (10-30%) Sediment – none Nutrients – none Metal – none Oil & Grease - none	nil	AUD 300 - AUD 500 Cleaned monthly by vacuum or remove the basket	Australia

Gross Pollutant Treatment Devices	Description	Catchment Size	Treatable Flow	Efficiency (pollutant capture)	Capital Cost (Supply and Install)	Maintenance Cost (per service)	Country
Grate Insert	– Grate inserts are typically be found in parking lots, alleys, and sloping streets. – Inserts installed in these basins mainly capture trash smaller than an inch due to the standardized grating spacing. – Inserts designed for curb opening basins are best suited for capturing larger debris like water bottles and plastics bags, as the opening under the curb may range from four to eight inches.	< 2 ha	nil	Litter – (10-30%) Sediment – none Nutrients – none Metal – none Oil & Grease - none	nil	RM 200 - AUD 300 Monthly by hand or vacuum truck	All Countries
Netting Devices							
Release Nets	Description – Release Nets are fitted to the end of stormwater pipes to capture gross pollutants. – When the net becomes full, a hydraulic drag mechanism releases the bag to prevent upstream flooding. – Release nets work on a similar principle to trash racks but at a smaller scale.	≤ 50 ha	nil	Gross Pollutant – (40- 70 %) Sediment: Course – (10-20%) Medium – none Fine – none Nutrients – none Metals – none Oil & Grease - none	AUD 2000 – AUD 10,000	Monthly or as required by small crane or Hand AUD 2,000 – AUD 10,000 per year	Australia

1. Source: 1. Orange County Stormwater Program Trash and Debris BMP Evaluation (2003), Report
    1. Stormwater Treatment Framework & Stormwater Quality Improvement Device Guidelines (SQID) (2003), Report
    2. Lake Macquarie Council (2003), Report
    3. Stormwater Gross Pollutants (1997), CRC Catchment Hydrology Report

2.2.2 SOURCES OF GROSS POLLUTANT

All forms of development and land use generate gross pollutants of one kind or another. In residential areas, the bulk of the volume of pollutant could be grass clippings with only small volumes of plastic, bottles and cans. Residential areas contribute pollutants as a result of household activities such as renovation works, painting, pet droppings, detergents and oils from car washing. Studies and logic indicate that a significant proportion of gross pollutants discharged to waterways are generated by residential land, as this type of development constitutes a significant proportion of the land use in most catchments (Land Development Guidelines, 2007).

In tourist attractions and general commercial and office areas, the type of pollutant is more likely to be floatable (i.e. cans, cigarette butts, paper, food wrappers, etc.) and motor vehicle generated pollutants (e.g. oils, brake linings, etc.). Industrial areas are more likely to generate gross pollutants such as polystyrene, wood particles, cardboard, and wrappings. These items, when discharged to waterways, are highly visible to the public.

Industrial sites are also more likely to generate spills of oil, chemicals and similar liquid contaminants, which are not generally trapped by physical gross pollutant control devices. Shopping Centre developments are more likely to concentrate pollutants related to food, packaging and motor vehicles (leak oil from parked vehicles, cars deposit brake linings, etc.). Park areas and rural developments are likely to generate volumes of organic matters (i.e. grass, leaves, etc.) and chemical pollutants associated with farming type land use. In assessing the source of a type of pollutant to be collected, consideration needs to be given to the potential change in pollutant source and type of pollutant which occur as a catchment develops or is redeveloped.

2.2.3 FACTORS AFFECTING AMOUNT OF GROSS POLLUTANT

Amounts of gross pollutants generated from different types of land use differ from one location to other location. It is important to acknowledge the factors affecting amounts of gross pollutant to manage gross pollutants, especially in urban waterways. The study by Allison and Chiew (1995) at a fully urbanized area in Coburg catchment shows the variability of the composition of gross pollutants with different types of land use.

Garden debris was found to contribute the highest constitution in terms of dry mass, followed by plastic. The same pattern of gross pollutant characteristic has been obtained for study in Auckland (Cornelius et al., 1994 and Island Care New Zealand Trust, 1996). Caltrans (2002) study performed at Orcas Avenue and Filmore Street found that vegetation has a higher percentage in terms of mass and volume compared to other types of gross pollutant. However, the study on gross pollutant characteristic in New York City

Streets found that plastic constitutes over 50% of the total mass percentage followed by metals and paper (England and Rushton, 2003).

Numerous factors affect the number of gross pollutants in urban waterways. RBF Consulting (2003) and Marais et al. (2001) concluded that various factors affecting gross pollutants deposition into waterways are listed in Table 2.1.

Table 2.2 Factors Affecting Amounts of Gross Pollutant (modified from RBF Consulting, 2003 and Marais et al., 2001)

Factors	Description
Type of Development	Generally commercial and industrial areas produce a higher amount of gross pollutant
Density of Development	Different density contribute a different load
Population	Permanent of transient residences
Rainfall Pattern	Intensity, stormwater runoff
Management Practices	Enforcement of street sweeping, garbage collection service, law enforcement on littering
Community Profile and Behavior	Income level, environmental awareness
Seasonal Variation	Longer dry periods usually accumulate more pollutants
Physical catchment characteristic	Size, slope, surface characteristic, type of vegetation
Drainage system	Size and geometry of inlet and pipe networks
Type of Industry	Some industries tend to produce more pollutants than others do
Type of Vegetation in the Catchment	Some trees have relatively large leaves which are slow to decompose
Refuse Removal Service	Availability of Street Sweeping
Legislation	The extent of legislation prohibiting or reducing waste, which is associated with the effectiveness of the policing of the legislation, and the level of the fines

2.2.4 MANAGEMENT OF GROSS POLLUTANTS

From observation, rapid urbanization is a major factor that contributes to the pollution of receiving waters, as it tends to change the environmental characteristics and affect the quality of the runoff. In Malaysia, the environmental problems associated with gross pollutants in urban waterways are recognized, and new approaches in the management of urban stormwater are made based on the concept of 'control at source' to control the stormwater quality and quantity (DID, 2012). Among the methods used to reduce gross pollutants from entering the rivers are described in the following subchapter.

- Non-structural Method

Non-structural methods approach focuses on public attitudes through monitoring and law enforcement. FHWA (2002) describes that non-structural Best Management Practices (BMPs) are an at-source approach to prevent and remove stormwater constituents load. Stormwater non-structural BMPs is one of the best cost-effective solutions to reduce constituents in stormwater runoff.

Public education hopes to increase public awareness of the impact of pollution on the environment. Law enforcement on littering also promotes the reduction of gross pollutants. Fundamental elements towards the successful implementation of non-structural methods are positive participation and involvement of individuals, communities and also government and private agencies.

Non-structural measures have been implemented in many countries around the world as litter management strategies or to reduce amounts of gross pollutants in waterways. In Australia, non-structural measures mentioned above are widely applied in its catchment's stormwater management. The Department of Environment and Conservation New South Wales (2004) has launched a stormwater education program – The Drain is Just for Rain in the year 1999 – 2004. The campaign is applied throughout New South Wales cities such as Waverly City, Dubbo City, Sydney, Rockdale City and Hornsby Shire. The ultimate outcome of this campaign demonstrates significant integration; the effective design and delivery of environmental education occurred best within a framework of integrated environmental management and sustainability.

The United States Environmental Protection Agency (USEPA) has also taken a serious approach to control stormwater pollution problem. The public involvement program implemented include facilitating opportunities for direct actions, educational, storm-drain marking or stream clean-up program. For example, Adopt-A-Stream programs are an excellent public outreach tool for municipalities to involve citizens of all ages and abilities. They are volunteer programs in which participants "adopt" a stream, creek, or river to study, clean up, monitor, protect, and restore. Through this activity, the adopting group or organization becomes the primary caretaker of that stretch of stream in the watershed. Another non-structural BMPs implemented in the United States of America is storm drain marking. The method used requires labeling storm drain inlets with plaques, tiles or painted pre-cast messages warning citizens not to dump pollutants into the drain.

In Malaysia, Drainage and Irrigation Department Malaysia has introduced a Public Outreach Programme (POP) under ROL Project. The River of Life Public Outreach Programme (ROL-POP) is a programme to foster partnerships and to improve attitudes and behaviors of target groups to reduce pollution in the Klang River, Malaysia. The main objective of ROL-POP is to generate evidential improvement in attitudes and behaviors of target groups within the Project Area towards river care and preservation to improve water quality and reduce pollution within the project area.

- Structural Method

The installation of trapping devices to trap and contain gross pollutants is a structural method to address gross pollutant problem in urban waterways. Mike et al., (2003) describe that at-source structural method to reduce the number of gross pollutants in the drainage system is a technique where infrastructure is installed at or near pollution sources. Structural measures installed to trap and screen gross pollutants, coarse sediments and non-point source that is carried through the catchment drainage system. Apart from trapping gross pollutants, some of the gross pollutant traps are also able to capture oil and grease. In Malaysia, DID Malaysia (2012) also introduce gross pollutant traps to remove litter, debris and coarse sediment from stormwater. However, improper maintenance of GPTs contributed to the clogging problem of the drainage system. Thus, it is essential to have an appropriate design and maintenance frequency of GPT to ensure the devices are working effectively.

2.3 SUMMARIES OF GROSS POLLUTANT STUDIES

2.3.1 LOS ANGELES GROSS POLLUTANT BASELINE MONITORING (LOS ANGELES COUNTY 2002, 2004A, 20014B)

The Los Angeles Country Department of Public Works performed gross pollutant baseline monitoring for five land uses in the Los Angeles River and Ballona Creek Watersheds between 2002 and 2004. High-density single-family residential, low-density single-family residential, commercial, industrial and open space parkland uses were monitored via the selection of storm drain inlets. For each land uses, a minimum of ten representatives' sites was sampled with each sampling site containing a minimum of five-catch basins fitted with storm drain inserts. In the Ballona Creek watershed, a total of 258 United Storm Water DrainPac storm drain inserts were installed and monitored the first year and an additional eight the second year (Los Angeles County, 2004a).

Individual drainage areas were determined for each storm drain inlet inside the City of Los Angeles by drainage maps and site visits. Outside of the City of Los Angeles, catch basin drainage area maps were created using the following approach (Los Angeles County 2002):

- From the as-built, locate flow lines along the streets that are captured by the targeted catch basin.
- Field check each site to confirm the catch basin location and flow paths
- Follow the flow lines up-stream to the starting point, the next catch basin or where their flow lines diverge to another catch basin, whichever is shortest. This is the limit of the flow captured by the targeted catch basin.
- If the targeted catch basin is located on one side of the street, the boundary line is drawn down the center of the street. This is because the streets are crowned (the center of the street is higher than the sides), which forces half to the other side.
- On private property, the boundary lines are drawn at a 45-degree angle starting from the street intersection toward the middle of the block.
- The sections of private property to be included in the drainage area are those bordering the streets where the targeted CBs are located.

In addition to monitoring storm drain inlets, one hydrodynamic separator with at least 5 catch basins upstream of the device fitted with storm drain inserts was monitored for each land use. The purpose of the hydrodynamic separators was to study the effectiveness of the upstream catch basin inserts during the study period (Los Angeles County 2002).

Four rain gauges were monitored in the Los Angeles River Watershed during the study period to assess storm intensity and totals. Debris within each of the hydrodynamic separators and storm drain inlets monitored was emptied within 72 hours of any of rain gauges receiving 0.25 inches of rain or higher. During dry weather, each device was cleaned every three months. In the Ballona Creek Watershed, one rain gauge was monitored to determine the monitoring event timing in that watershed.

For the first year of monitoring, there were nine wet weather cleanouts and one dry weather cleanout in both watersheds (Los Angeles County 2004a). During the second year, there were eight wet weather cleanouts in the Los Angeles River watershed and six in the Ballona Creek watershed (Los Angeles County 2004b). Each storm drain and hydrodynamic separators was cleaned, and debris was individually bagged and tagged. The contents were then sorted by separating sediment and vegetation from the man–made gross pollutant, and both categories of material were weighed and volume recorded. The material was not dried prior to measurement.

Commercial land use was the highest contributor of man–made gross pollutant per acre of the drainage area (by weight) in the Ballona Creek watershed during both years of the baseline study. The loading rate from commercial land use was at least twice as high as all other land use types in the Ballona Creek watershed. In the Los Angeles River watershed, industrial land use had the highest rates in both years, while commercial land use had similar rates during the first year of sampling (Los Angeles County 2005).

For both watersheds, the first seasonal flush was responsible for a sizeable portion of the gross pollutant collected during the study. In 2002–2003, the first storm contributed approximately 40% of the gross pollutant collected during the entire year (42.2% for the Los Angeles River watershed and 36.3% for the Ballona Creek watershed). During the second year, the first storm of the season contributed most of the gross pollutant (23.6%) collected in the Los Angeles River watershed during the entire year, but in the Ballona Creek watershed, the second storm contributed the grossest pollutant (28.1%) for that year (Los Angeles County 2004a).

2.3.2 SPRINGS AND ROBINSON CANAL TRAP STUDIES (ARMITAGE ET AL. 1998)

This study was conducted to estimate the quantity and types of gross pollutant coming off a catchment (299 ha) in Springs, South Africa that drained to a single outfall. A gross pollutant removal structure, capable of handling 7.5 cubic meters per second (m3/s) before bypass occurred, was installed at the outfall. The structure was designed to capture particles larger 20mm. From the 14 samples collected, an average of 3.3 cubic meters (m3) of the gross pollutant was associated with each storm event. The normalized data showed that the average gross pollutant loading rate was 4.9 m3/ha·yr. Roughly 87% of the gross pollutant generated onto the streets was removed by street sweeping.

The Robinson Canal Trap Study was designed to assess gross pollutant loadings in Johannesburg, South Africa. Similar to the Spring Study described above, a single gross pollutant removal structure that could handle 15 m3/s of flow before bypass and screened out particles larger than 20mm was installed on the outfall of a catchment. The trapped material was equal parts sediment, suspended debris and floatables. During the first seasonal flush, more than 150 plastic bags were collected. Results indicate that gross pollutant is washed into the stormwater drainage system at a rate of 0.50 m3/ha·yr.

Based on data from the Springs and Robinson Canal studies, the authors developed the following equation to predict the total annual gross pollutant and vegetation loads to water bodies in South Africa from the stormwater drainage system:

$$T = \Sigma fsci \cdot (Vi+Bi) \cdot Ai \qquad \textit{Equation 2.1}$$

where:

T = total gross pollutant load to water bodies
fsci = street cleaning factor for each land use (dimensionless)
(This factor relates the anticipated efficiency of street cleaning. If there is no difference, fsci = 1).

$$fsci = (1 - \eta design) / (1 - \eta data) \qquad \textit{Equation 2.2}$$

where:

ηdesign = anticipated efficiency of street cleaning (fraction)
ηdata = efficiency of street cleaning during data collection (fraction)
(e.g. an improvement in removal due to street sweeping from say, 15% of the total load to 75% of the total load will indicate fsci = 0.294)
Vi = vegetation generation rate for each land use (m3/ha·yr)
(Varies from 0.0 m3/ha·yr for poorly vegetated areas too, say 0.6m3/ha·yr for densely vegetated areas)
Bi = basic gross pollutant generation rate for each land use (m3/ha·yr)
Ai = area of each land use (ha)

2.3.3 CAPE TOWN CASE STUDY (MARAIS ET AL. 2004)

This study aimed to measure the amount and type of gross pollutant entering the drainage system from different urban land uses and demographics in Cape Town, South Africa. Nine catchments representing a range of different land uses and socio-economic profiles were equipped with at least one catch pit traps (similar to catch basin inserts). Catch pit traps were cleaned, on average, once a month and collected material were weighed and sorted.

Data from this study, shown in Figure 2.1, found that for many sites, sediment and vegetation loads are sizeable portions of the material captured. Excluding sediment and vegetation loads, plastic was the largest portion of the total gross pollutant load. Loads are highest in low-income residential areas, followed by mixed commercial and industrial areas. A decrease in gross pollutant was observed with increasing income, though population density in these areas decreases, which increases the per capita gross pollutant load. The catch pit traps collected one to three % of gross pollutant removed when material removed by street sweeping is considered as well.

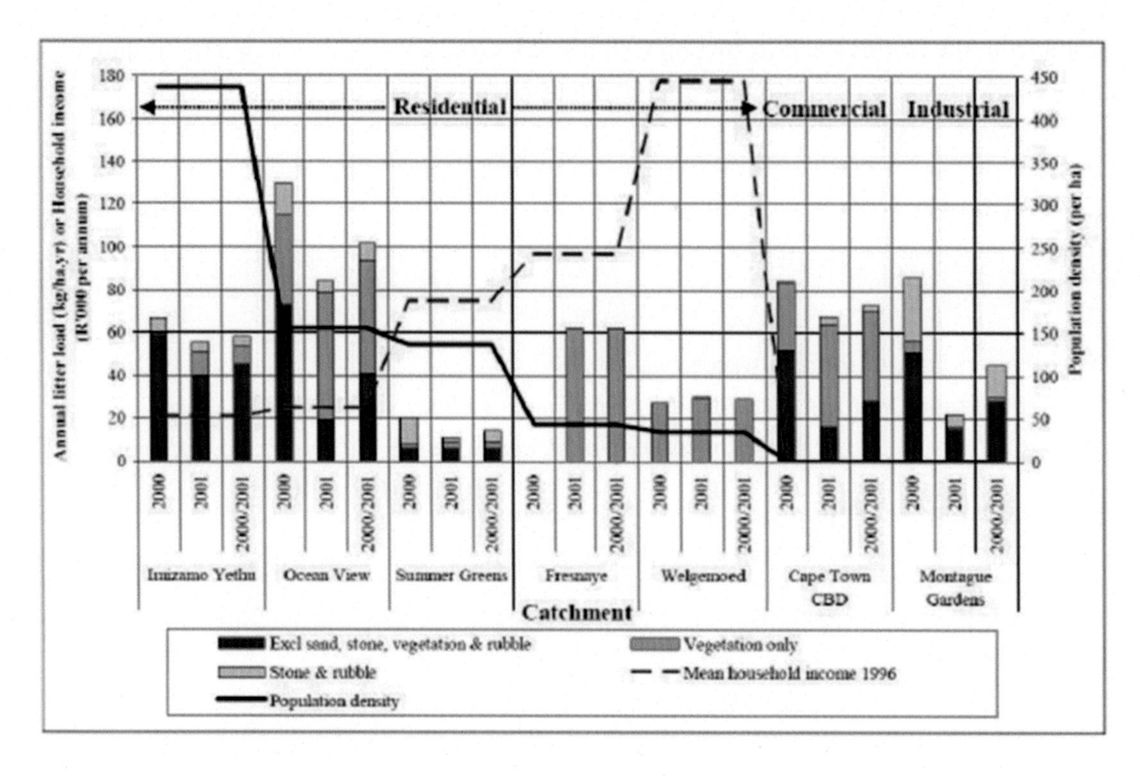

Figure 2.1 Summary of results from Cape Town (Marais et al. 2004) for annual gross pollutant load against the population and mean household income.

2.3.4 *Highway Event Mean Concentrations (Kim et al. 2004)*

Conducted in Los Angeles, this study estimated gross pollutant loads and determined the event mean concentration (EMC) of gross pollutant in stormwater from highways. Mesh bags were placed over stormwater discharge pipes and collected at the half-hour, one-hour and two-hour intervals after runoff began. A fourth bag was collected at the end of the storm event. Vegetation was separated from a gross pollutant, and the gross pollutant was separated into biodegradable and non-biodegradable materials. Wet and dry weight and volumes were measured. Results were correlated to average daily traffic, antecedent dry days, and total rainfall. The study found that 90 % of the material collected was vegetation. The equation below was also developed to predict the gross pollutant event mean concentration (EMC) based on total rainfall (TR) and antecedent dry days (ADD).

EMCgross pollutant (g/L) = ε(ADD)a(TR)b *Equation 2.3*

where:

ADD = antecedent dry days (days)
TR = Total rainfall (cm)
ε, a, b = fitting parameters defined in Table 2.2.

Overall, the event mean concentrations decreased with total rainfall but increased with antecedent dry days. Additionally, fitting parameters were also derived based on the results of this study (Table 2.2).

Table 2.3 Event mean concentration fitting parameters

Parameter	ε	a	b
Wet gross pollutant	0.0239	0.206	-0.408
Wet vegetation	0.02056	0.215	-0.387
Wet Gross pollutant	0.0016	0.360	-0.683
Dry gross pollutant	0.00095	0.354	-0.694
Dry biodegradable gross pollutant	0.00061	0.245	-1.034
Dry non-biodegradable gross pollutant	0.00032	0.336	-0.408

2.3.5 NEW YORK CITY STREET SWEEPING STUDY (NEWMAN ET AL. 1996)

In order to assess the ability of manual sweeping of street curb areas to reduce gross pollutant on New York City streets, a four-month study was conducted in the summer of 1993. Over 15-block faces in New York City, sweeping was conducted using the following methods/frequencies:

1. Baseline – standard mechanical street sweeping two (2) times per week.
2. Level 1 Enhancement–Baseline mechanical sweeping plus manual sweeping once a day six (6) times a week.
3. Level 2 Enhancement – Baseline sweeping plus manual sweeping twice a day six (6) times a week.

Once a week during the study period, all litter was collected from the 15 block faces with a push broom ignoring vegetation, sediment and items under parked vehicles. The collected material was weighed, sorted into 13 categories. Item counts, total weights, wetness, and loosely packed density were also measured.

Results indicated that there was a significant decrease in gross pollutant on streets when manual sweeping was implemented. The marginal benefit in gross pollutant reduction (400-65%) was most pronounced when Level 1 cleaning (sweeping an additional six times per week) was implemented. A much smaller marginal benefit (5-10%) was observed when Level 2 cleaning was implemented. Due to the relatively low amounts of gross pollutant in low-density residential areas, enhancing sweeping to Levels 1 and 2 had no observable effect. In contrast, enhanced sweeping had the most impact on streets near vacant lots where the highest levels of gross pollutant were observed.

An additional finding of this study was that block faces located in Business Improvement Districts (BIDs) had 75 % less gross pollutant on streets than those without BIDs. Street and sidewalk cleaning by private contractors hired by BIDs were assumed to have caused a decrease in gross pollutant in these areas.

2.3.6 New South Wales Gross Pollutant Trap Assessment (Wong and Walker 2002)

The New South Wales Environmental Protection Authority commissioned a report to develop a methodology to assess the performance of gross pollutant traps installed by local government agencies. Though the purpose of the report was to assess the cost and capture performance of the traps, the authors included the data from the cleanouts of 28 devices by local government agencies. Of the 28 devices considered, 12 were continuous deflective separation (CDS) units, which are considered full gross pollutant capture devices by the San Francisco Bay Water Board. These CDS units were located in residential areas and mixed residential/commercial areas. The cleanouts occurred between March 1999 and August 2000, with some devices cleaned only once and another clean up to eight times. At the time of the cleanout, the material was weighed, and the percentage of sediment, gross pollutant and organic material was estimated.

There was little difference in the composition of material removed from the 12 CDS units for the four seasons. On average, 14 % of the material removed, by weight, was a gross pollutant, 30% was sediment and 56 % was organic material for all seasons and land uses. However, winter and fall cleanouts had slightly higher average percentages of gross pollutant (16 and 15%, respectively) than spring and summer cleanouts (with 13% gross pollutant each). In terms of land use, the residential/commercial areas had a much higher percentage of gross pollutant (23 %), than the purely residential areas (the majority of sites) that averaged 13 % total pollutant.

2.3.7 Coburg Stormwater Gross Pollutants (Allison et al. 1998a, Allison and Chiew 1995)

The Cooperative Research Center for Catchment Hydrology in Australia conducted a study in 50 ha mixed residential/industrial catchment in Coburg, Australia (eight kilometers north of Melbourne) to determine the quantities and characteristics of gross pollutants in the stormwater system associated with rainfall events and assess the performance of a hydrodynamic separator, accurately, a continuous deflective separation (CDS) unit, and catch basin inserts known as side entry pit traps (SEPTs). The catchment is 35% commercial and 65% residential.

The first part of the study involved material removed from the CDS after single storm events from May to August 1996. The unit was cleaned after at least 36 hours of dry weather following the storm (to ensure low inflows). A swimming pool leaf-scoop removed floatables while sump material was removed manually using a rake and vacuum after water was pumped out. Floatable and sump material was drained, categorized, and oven-dried. The weights and volumes of the categories were then measured. For the second part of the study, 192 catch basin inserts were installed in the catchment. Every two or four weeks, the catch basin inserts and hydrodynamic separator were a cleanout on the same day. Material from all devices was handled in the same manner as the first part of the study.

Based on the CDS cleanouts, it was found that approximately 7.6 kilograms per hectare (kg/ha) of gross pollutant is washed off the catchment each year and that there is a logarithmic relationship between the litter load and storm event rainfall and runoff, as shown in Figure 2.2 and Figure 2.3, respectively. Additionally, the density of gross pollutant material is lower than organic material, and approximately four times as much gross pollutant is suspended or sinkable (i.e. CDS sump material) than floatable (only 20% of the gross pollutant load floats) as seen in Figure 2.4. However, since the material in the CDS sump was retained for a long time between the storm and the cleanout, the sump density may have been higher due to the settling of floatables.

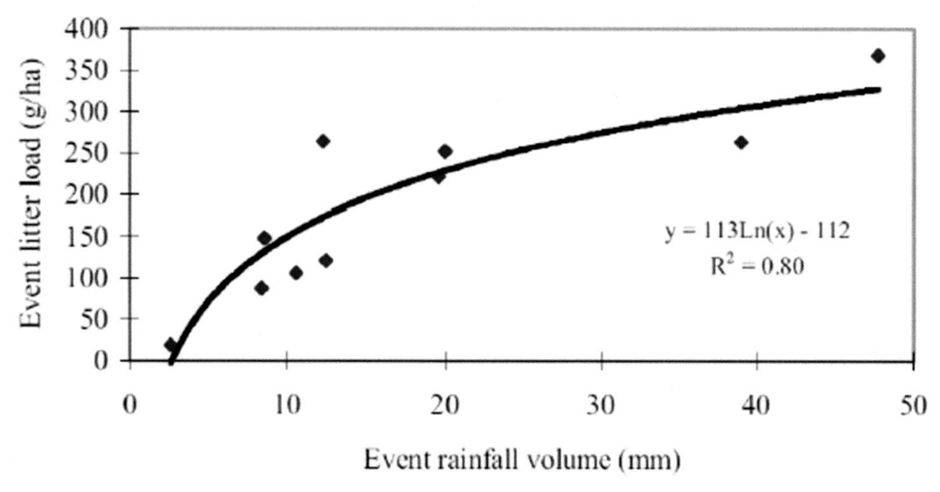

Figure 2.2 Rainfall versus gross pollutant load for 10 CDS cleanouts (Allison et al. 1998a, Allison and Chiew 1995)

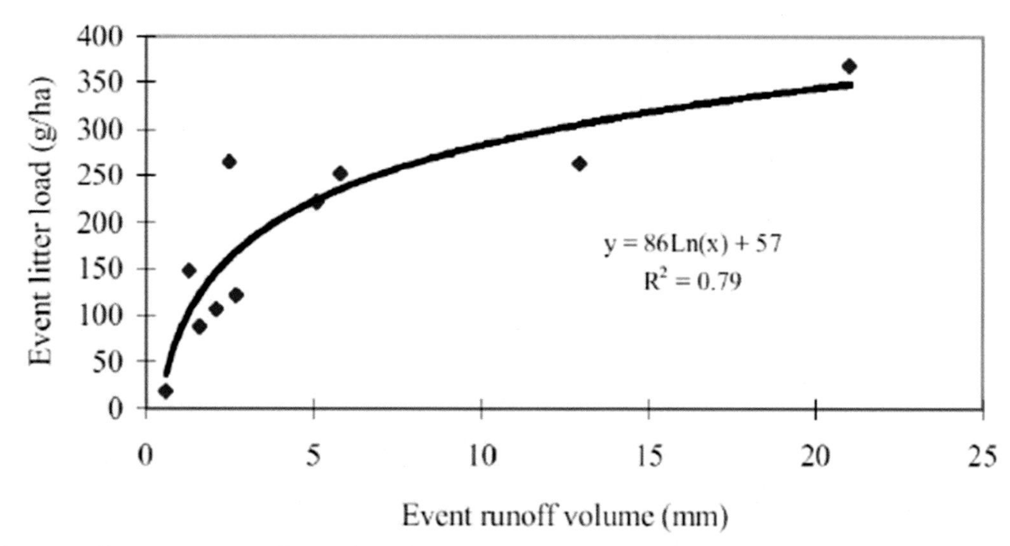

Figure 2.3 Runoff versus gross pollutant load for 10 CDS cleanouts (Allison et al. 1998a, Allison and Chiew 1995)

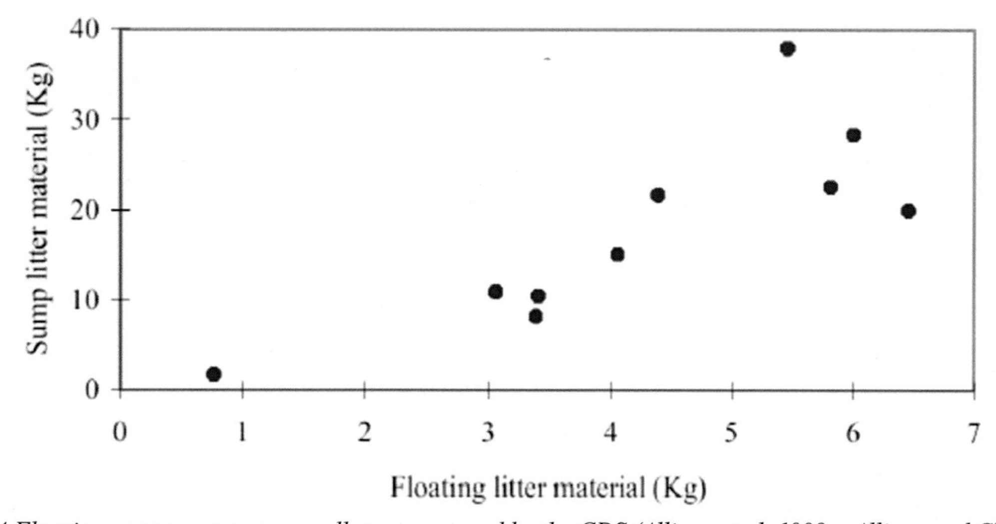

Figure 2.4 Floating versus sump gross pollutant captured by the CDS (Allison et al. 1998a, Allison and Chiew 1995)

From the second part of the study, it was found that fast-food consumers and smokers were a significant source of gross pollutant in urban streams. The CDS units were determined to be approximately 100 % efficient, and SEPTs capture about 80 % of the gross pollutant load.

2.3.8 *Decision Support System (Allison et al. 1998b)*

This paper describes the "Decision-Support-System" (DSS) used to select the appropriate gross pollutant trap for a particular urban area. Users input land-use composition and rainfall, and the model estimates the typical load (i.e., the load generated). Incorporating street sweeping behavior (supplied by the user for the area of interest) yields an actual load. The model outputs include the dry mass retained by the trap, the receiving water load (in dry mass) and the costs.

2.4 GROSS POLLUTANT LOADING RATES

A comparison of gross pollutant loading rates developed through the studies summarized above is illustrated in Table 2.2. Loading rates are stratified by land use, where available. Studies with no data regarding loading by land use are excluded. Based on this comparison, it appears that gross pollutant loading rates differ between studies. Based on his extensive experience in developing gross pollutant loading rates in South Africa, Armitage (2007) states that there is an enormous variation in measured gross pollutant loading due in part to the lack of uniformity in reporting data with respect to moisture content, density and composition (including or excluding leaves and sediment).

Based on these tables, there are a few conclusions that can be drawn:

- Commercial areas generally have the highest gross pollutant load rates among all land use categories;
- Residential areas consistently have the lowest per-unit gross pollutant loads of all land uses monitored, with the exception of the low-income areas of Cape Town (Marias et al. 2004);
- Vegetation or garden debris can account for a sizeable portion of gross pollutants, especially in high-income residential areas; and,
- Excluding vegetation and sediments, plastic and paper are the largest portions of urban gross pollutant, particularly in commercial and low-income residential areas.

Table 2.4 A summary gross pollutant loading rates as calculated by case studies reviewed (EOA, 2011)

Study	Units	Industrial	Commercial	Residential		Park Land	All	Gross Pollutant Density (kg/m3)
				Low Density	High Density			
Los Angeles River Watershed, Los Angeles County – USA (Los Angeles County 2004b)	kg/ha.yr	16.69	12.71	0.71	3.28	3.26	6.73	139.8
Ballona Creek Watershed, Los Angeles County – USA (Los Angeles County 2004b)		2.47	9.58	3.12	3.40	3.10	3.78	
Cape Town - South Africa (Marais et al. 2004)	kg/ha.yr	28[a]	22-59[b]	41-45[c]/6d/0[e]		-	-	-
Springs – South Africa (Armitage et al. 1998)	kg/ha.yr (m3/ha.yr)	550 (5.8)		-	96	-	470 (4.9)	Dry density = 95 – 150
Johannesburg – South Africa Robinson Canal trap (Armitage et al. 1998)	kg/ha.yr (m3/ha.yr)	-		-	-	-	48 (0.05)	Dry density = 95
Auckland – New Zealand (Cornelius et al. 1994)	kg/ha.yr (m3/ha.yr)	0.880 (0.009)	1.350 (0.014)	-	0.530 (0.006)	-	-	-
Coburg – Australia (Allison and Chiew 1995)	kg/ha.yr	7.4[a]	17	-	4.0	-	6.0	Wet density = 260 (including leaves)

[a] Light industrial [b] Mixed commercial and industrial [c] Low income [d] Moderate income [e] High income [f] Includes leaves

2.5 CONCEPTUAL MODEL FOR GROSS POLLUTANT

Gross pollutant loads in urbanized areas are dependent upon:

- Gross Pollutant Generation – the rate at which gross pollutant is generated (i.e., deposited onto the urban landscape); and,
- Gross Pollutant Interception – the degree to which gross pollutant is intercepted through control measures (e.g., street sweeping) before entering drainage system.

While many anthropogenic factors likely affect the rate at which gross pollutant is generated in urbanized areas, based on the studies conducted to-date the following three (3) factors have been demonstrated as influential:

- Land Use – Residential areas often produce less gross pollutant than industrial and commercial areas (Los Angeles County 2004, Marais et al.2004, Allison and Chiew 1995, Cornelius et al. 1994);
- Economic Profile – Gross pollutant generation rates appear to have an inverse relationship with income (Marais et al. 2004); and,
- Population Density – A higher density of people will produce more gross pollutant even if the per capita generation rate is lower (Marais et al. 2004).

The effectiveness of gross pollutant interception prior to reaching a drainage system is also dependent on the following factors:

- Street Sweeping Effectiveness– Higher frequencies and efficiencies reduce the likelihood of gross pollutant available for transport to drainage system during storm events (Lippner et al. 2001); and,
- Manual Pickup – Augmenting street sweeping with manual gross pollutant pick-up can significantly decrease the gross pollutant on streets available for transport to drainage system (Newman et al. 1996).

In addition to these anthropogenic factors that affect gross pollutant generation and interception, the literature review suggests that the following two (2) natural environmental factors are also essential to consider when establishing gross pollutant loads:

- Rainfall (Totals and Intensity) – an increase of gross pollutant loads in stormwater have been documented with an increase in rainfall (Kim et al. 2004, Allison et al. 1997, and Lewis 2002) because larger runoff events are capable of carrying more gross pollutant to drainage system; and,
- Antecedent Dry Days – due to the build-up of the gross pollutant on streets, gross pollutant loads in stormwater typically increase as the number of days between rainfall events increases (Kim et al., 2004).

Considering these five (5) anthropogenic and two (2) natural factors which these seven factors are significant, there are other noteworthy factors that are difficult to quantify, but will also potentially affect gross pollutant loads. Some of these factors include legislation and enforcement, public education and awareness along with a community concern for the environment, a sense of ownership of the area, whether or not the area is already dirty, and availability of gross pollutant receptacles. A conceptual model for gross pollutant loads from the drainage system was developed. The conceptual model attempts to control the most important and quantifiable factors that influence gross pollutant loading rates from the drainage system. The conceptual model allows for testing the importance of these factors through monitoring of gross pollutant loads from specific GPTs in ROL project. In this study, the same concept will be adopted (EOA, 2011). The proposed concept is shown in Figure 2.5.

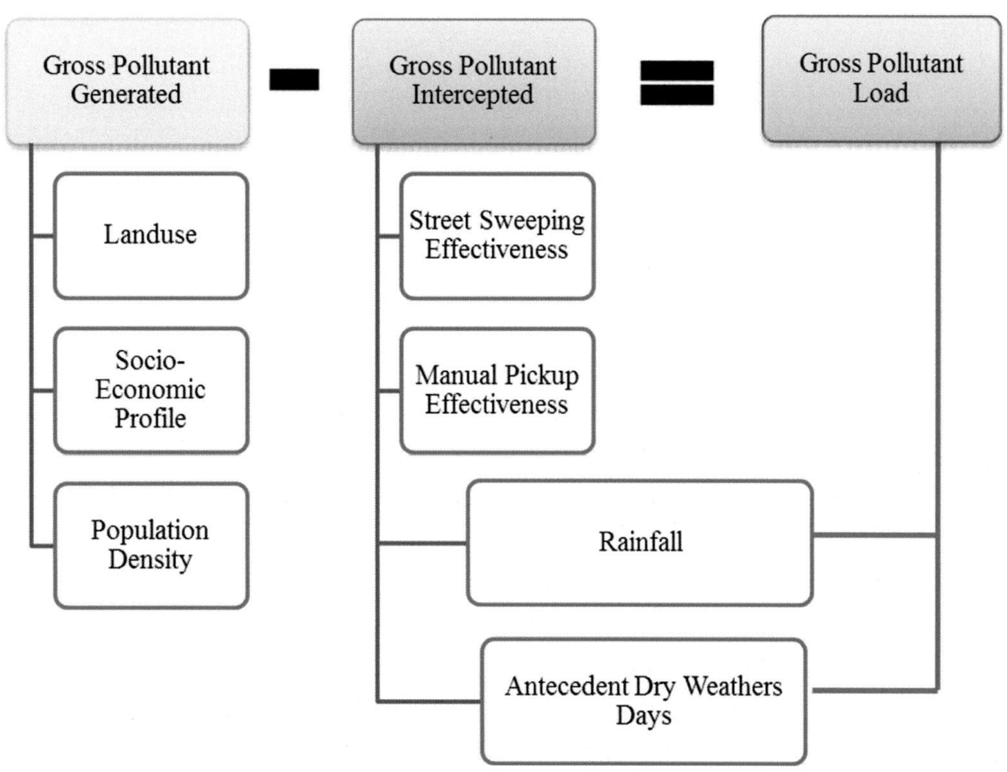

Figure 2.5 Proposal model of gross pollutant loads for ROL Project, including proposed influencing factors (adapted EOA, 2011)

2.6 GROSS POLLUTANT TRAPS

GPTs are a device that is purposely designed to remove litter, debris and sediment from stormwater. It is fixed at the downstream end of the drainage system which is before entering the waterway such as a river, pond or wetland. The traps also serve the purpose of filtering floatable oil, dirt, chemicals, bacteria and other small pollutants provided that they are designed correctly.

Gross pollutant traps combined the mechanism of total solids interception and retention. This mechanism utilizes the energy coming from the inflow to separate floatable materials with non-floatable materials. For examples, sedimentation tank is used to settle the non-floatable material such as sediments, trash rack applying the intercept mechanism by gross intercept solids from stormwater and CDS utilized both mechanisms of interception and retention.

2.6.1 Type of GPTs

There is a very wide range of devices for the treatment of gross solids. Selection of suitable devices depends on many factors, including catchment size, pollutant load, the type of drainage system and cost (DID, 2012). Table 2.4 presents the classification of the kinds of gross pollutant traps that could be used in Malaysia.

Table 2.5 Gross Pollutant Traps Categories (DID, 2012)

Class	Functions	Catchment Area	Installation
Type 1 - Floating Debris Traps (Booms) - Trash Racks & Litter Control Devices	Litter capture on permanent water bodies	>200 ha	Proprietary and purposely built-on-line
	Litter and sediment capture in the existing pit	2 – 40 ha	Purposely built from modular components – online
Type 2 – Sediment Basin and Trash Racks (SBTR) Traps	Sediment and litter capture on drainage conveyance	5 – 200 ha	Proprietary and purposely built-on-line of off-line
Type 3 – Sediment, Oil and Grease Interceptor	Oil, grease and sediment capture on drainage conveyance	< 40 ha	Proprietary – off-line

2.7 PUBLIC OUTREACH PROGRAM (POP)

The River of Life Public Outreach Program (ROL-POP) is a program to foster stakeholder partnerships and to improve attitudes and behaviors of target groups to reduce pollution in the Klang River, Malaysia. It is a component of the ROL initiative mooted by the Department of Irrigation and Drainage Malaysia. This section/component is jointly implemented by ERE consult and Global Environment Centre ((ROL-POP), 2012 - 2017).

The Project site covers the 10 km stretch of Klang River representing about 40.4 km^2 catchment area (from Klang Gates Dam to the confluence of Sg. Klang and Sg. Ampang). About 80% of the Project site lies within municipality area of Ampang Jaya Municipal Council while the other 20% lies within the municipality area of Kuala Lumpur City Hall. The land use of the Project site is predominantly urban. Residential land use accounts for 38% of the area while commercial and industrial areas account for 7% and 5% respectively. The main tributaries of Sg. Klang within the Project site is Sg. Kemensah, Sg. Sering and Sg. Gisir. The population within the Project site is estimated to be about 146,000.

The main objective of ROL-POP is to improve attitudes and behaviors of target groups to enhance the quality of water and reduce pollution within the project area. It will address pollution management by

i. **Educating the public on common do's and don'ts as well as other skills to preserve the rivers**

ii. **Promoting a sense of ownership towards the river**

iii. **Initiating long-term and sustainable change in behaviour towards preserving the river.**

Target Groups Targeting the upper reaches of the Klang River basin, ROLL POP engages with five specific target groups:

i. **General public,**

ii. **Targeted local communities,**

iii. **Schools,**

iv. **Restaurants and food courts,**

v. **Property developers and industries in various ways.**

2.7.1 Relationship between Behaviour of People with GPT

According to ROL POP final report at Upper Klang, survey from March 2014 to May 2014 about general public's perception on river care, majority of the respondent (54%) were still aware of the location on the river nearest to their homes and most of them still believes that the rives near their homes to be in a poor state. However, most of the respondents (50%) were not sure where the drains went, while only 17.9% of the respondents believed the drained led to rivers. On the other hands, 23.7 of the respondent believe that the drains flowed into the sewer. From these points of view, we can see that with this lower awareness of the respondent, it will attribute to the accumulation of discarded waste and lead to increasing wet load in GPT.

Consequently, 43% of respondents did not think that they harm the river. But on another hand, 32% responded admitted that their activities impacts with the most significant causes identified as rubbish from the household (25.5%) and laundry wastewater 23.6%. In the same way, 25% of the respondent was unsure what impact they have to river. It appears that the behaviour of the respondent will affect the increasing of wet load on nearest GPT.

Based from the previous baseline survey, over the past two years 2010 to 2012, many respondents have taken part in some outreach activities like 'gotong-royong' (41.9%) and river clearing programmers (7.4%) to shows their environmental concern. Otherwise, 39.2% did not take part in any such activities. Meantime, in 2014-2014, many respondents still took part in reach activities such as 38.1% participated in 'gotong-royong' and 3R (Reduce, reuse, recycle) campaign (7.5%). As many as 13.8% of respondents have involved in tree planting, 5% in the river cleaning programme and some have engaged in training, briefing and river canal.

It appears that the general public has participated in more outreach activities than before, with 22.2% not participating in any activities, significantly less than the previous percentage of 39.2%. However, the majority of respondents still did not feel that their efforts had much impact; with the most respondents giving a rating of medium (36.4%), no effect (30.6%) and low (20.9%). To some extent, with the willingness of the respondent to participate in the outreach programme can reduce the wet load on GPT.

2.8 LIFE CYCLE COST

AS/NZS 4536:1999 defines several phases in the lifecycle of a product or asset, which represent 'cost elements' which include acquisition, use and maintenance, renewal and adaption and disposal. A conceptual diagram of these phases in the life cycle and cost elements and its associated potential elements is shown in Figure 2.6. The diagram is a modified diagram from Standards Australia (1999).

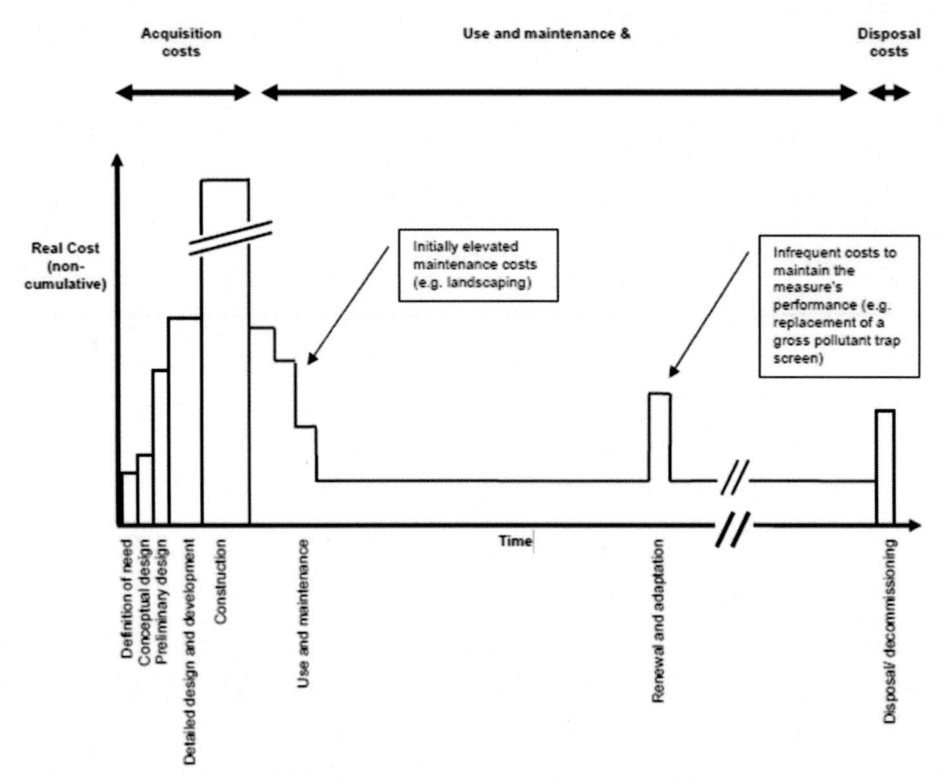

Figure 2.6 Phases in the Life Cycle of a Stormwater Quality Management Measure and Potentially Associated Cost (Taylor, 2003)

2.8.1 Installation Cost

Overall installation cost should include the cost of the trapping system and installation of a GPT. The construction cost of a GPT depends on the site condition in terms of the type of soil, access to the site and also groundwater condition. Mainly, an estimation of the installation cost requires consideration of the following issue:

i. **Design flow rate**

ii. **Size and configuration of the trap**

iii. **Hydraulic impedance and the requirement for operation, and**

iv. **Safety and other construction issues**

2.8.2 Use and Maintenance Cost

Maintenance costs are more difficult (but are sometimes the most critical variable) to estimate that the installation costs. Variances of the techniques used, the amount of material removed, and the unknown nature of the pollutants exported from a catchment has to be considered. One crucial step is to check with previous installations by contacting current owners of the litter trap and asking their annual costs.

Disposal costs are depending on whether the collected material is retained in wet or dry conditions. Handling of wet material is more expensive and require sealed handling vehicles.

i. **Is the material in a wet or dry condition, and what cost implications are there?**

ii. **Are there particular hazardous materials collected and require special disposal requirements (e.g. contaminated waste -what cost implications are there?)**

iii. **What is the expected load of material, and what are the likely disposal costs?**

Taylor (2005) stated that NSW EPA had developed a spreadsheet that provides some information on an approximate unit price and typical maintenance cost of proprietary GPTs, as shown in Table 2.5. Hornsby Shire Council (2003) has done monitoring on several types of GPTs. From this study, the council has obtained the maintenance cost of trash rack, large trash basket, proprietary device and side entry pit inserts, as shown in Table 2.6.

Table 2.6 Capital Cost and Maintenance cost of Proprietary GPTs (Taylor, 2005)

Type of GPT	Capital Cost (AUD)	Maintenance Cost (AUD)
– Rocla Downstream Defender	– 12,000 to 36,000	– 20/ha/month
– Stream Guard – catch basin insert	– 290	– 200 p.a.
– Stream Guard – passive skimmer	– 60	– 200 p.a.
– Enviropod	– 400 – 620	– 200 p.a.

Type of GPT	Capital Cost (AUD)	Maintenance Cost (AUD)
– Ecosol RSF 100	– 430 - 903	– 200 p.a.
– CSR Humes Humeceptor	– 10,000 – 50,000	– 20/ha/month
– Rocla Cleansall	– 20,000 – 150,000	– 14,400
– Ecosol RSF 1000	– 4,000 – 12,000	– 12/ha/month
– Baramy	– 15,000 – 40,000	– 12/ha/month
– CSR Humegard	– 18,000 – 51,000	– 14,400 p.a

Table 2.7 Gross Pollutant Trap Maintenance Cost from July 2001 to June 2003 (Collins et al., 2003)

Device	Average Capital Cost AUD/ha	Average Clean Cost AUD	Average Annual Maintenance Cost AUD/yr	Average Annual Maintenance Cost AUD/ha
Trash Rack (n=18)	972	210	2346	39
Large Trash Basket (n=9) Capacity = 0.15m3	2117	86	708	42
Proprietary Device (n=35)	5160	255	765	196
Side Entry Pit Inserts (n=77)	2800	25	300	1200

The analysis on GPTs trapping efficiency of a device can be defined as the proportion of the total mass of gross pollutants transported by stormwater that is retained by the trap. A trap with low trapping efficiency means that a higher proportion of the gross pollutants transported by the stormwater passing the trap and reaching downstream waters (Wong and Walker, 2002).

The selection of the most appropriate trapping devices must be relevant to the cost of installation and maintenance of the proposed GPT. A life-cycle cost is recommended for assessing the overall cost of GPT. This approach considers all costing aspect of GPT, which include the capital cost of the system, maintenance, servicing and spoil disposal cost over the system's lifespan. Sidek LM., 2004 defines the life-cycle cost for gross pollutant trap as given by Equation 2.4.

LCC (RM)=Construction Cost + (n ×Maintenance Cost) + (n ×Management Cost)

Equation 2.4

where;

n is the project duration in years

An Equivalent Annual Cost (EAC) of GPT is estimated by dividing the LCC of the project with the project duration as given in Equation 2.5

EAC (RM/yr) = LCC (RM) / duration (yr) *Equation 2.5*

A checklist for determining the life cycle costs of GPTs is listed in Table 2.7 (WSUD, 2010):

Table 2.8 LCC Checklist (WSUD, 2010)

Installation	**Y/N**
1. Does the trap satisfy:	
Flow Rate	
Available space constraints	
Hydraulic and flooding issues	
Other concerns (e.g. safety and aesthetics)	
If no to any of the above, then go no further	
2. Trap costs	
3. Installation cost	
4. Others (rock excavation, lid loading, the access road for maintenance etc.)	
Maintenance	
Annual maintenance cost	
Cost of any special maintenance equipment	
The expected cost of disposal	
Life cycle cost	
Estimated project duration (in years)	

2.8.3 Cost-Effectiveness Ratio (CER)

In other to quantify the value for money for each device, the "Cost-Effectiveness Ratio (CER)" analytical technique links trap capture outcomes with GPTs input costs. CER provides a simple tool to assess management options for pollution trap operations. CER was defined by Brisbane City Council (2002) as:

CER = LCC (RM/yr) / Pollutant Removal Efficiency (kg/ha/yr) *Equation 2.5* where,

PRE is the pollutant removal efficiency for the GPT (kg/ha/yr)

The Pollutant Removal Efficiency is calculated through a simple calculation of:

PRE = WPR / CA *Equation 2.6*

where,

PRE is pollutant removal efficiency (kg/ha/yr)
WPR is weight f pollutants removed (kg/yr)
CA is catchment area (ha)

These equations provided a maintenance effectiveness ranking that could be used in considering future GPT selection. Generally, devices with the lowest CER are preferred, meaning that they captured more pollution and were not excessively priced to clean.

2.9 GEOGRAPHICAL INFORMATION SYSTEM (GIS)

2.9.1 Introduction

GIS plays an important role in achieving this task when monitoring and analyzing the adverse environmental impact on water bodies, such as lakes, rivers, and bays. For example, a GIS facilitates an evaluation of pollution sources by generating reports and managing data about polluters, results of measurements, reference materials providing a classification of hazardous groups, and concentrations of hazardous substances in a specific river or an entire aquatic system. A GIS also provides the tools to identify the most hazardous contaminants about environmental regulations and contribute to effective decision making to ensure that natural resources are preserved and utilized correctly.

2.9.2 The use of GIS

The best way of carrying out the trap selection procedure is with the aid of a Geographical Information System (GIS) (Armitage, 2006). This is readily illustrated by way of a study that was carried out by Jeffares & Green Incorporated in consortium with Neil Armitage Consulting cc for the City of Cape Town, South Africa (Wise & Armitage, 2002; Marais & Armitage, 2003).

The GIS data sets that were relevant to this study (for example, topography, drainage, and land use) were obtained from the Cape Metropolitan Council Administration: Catchment Management Department. These data sets were used to develop a map of the two catchments.

2.9.3 Web-based GIS

Web-based GIS is a system that includes elements of user-friendliness, enabled data access, and enterprise data management to ensure that analysis and other functions done accurately and quickly.

Geospatial technology and data extend the capabilities of GPT's management and maintenance in a number of fundamental ways, including:

i. **Enhancing spatial context**

Along with time, location is a fundamental reference point for countless human activities. Location data provides a useful context that makes other geospatial or location related data more meaningful. For example, the capability to visualize the locations of recent failures on a map provides the context needed to spot a trend that could be missed by looking at a table of data.

ii. **Improving measurement capabilities**

Geospatial data help us understand the physical relationships among assets, such as the distance between them for routing purposes. This data also makes it possible to judge proximity, such as determining whether an asset is within a certain distance or something.

iii. **Extending modeling options**

Analyzing and visualizing geospatial patterns help identify trends and predict future events with greater accuracy.

iv. **Gaining deeper knowledge about GPT's locations**

Many organizations with widely dispersed assets, such as municipal water utilities, electric and gas distribution utilities, and departments of transportation, find it useful to track the locations of assets over time. GIS provides a robust framework for managing these types of data to better support asset management and maintenance activities.

v. **Improving visualization capabilities**

Visually displaying location data on maps is the most familiar, and often the most valuable, use of geospatial technology. Applying this capability to strategic assets has a vast range of implications for improving business performance.

2.9.4 Main GIS Components

GIS components should include GIS Application Server, GIS desktop, GIS spatial database management and mobile apps for data collection.

i. ArcGIS Server

ArcGIS Server is the foundation for distributing GIS data and applications on the Internet. By providing a common platform for exchanging and sharing GIS resources and provides unique opportunities to leverage data from within the organization and to integrate information from other agencies. Key features include data integration, standards-based communication, and the Internet-enabling technology, easy-to-use framework, a multi-tier architecture, support for a wide range of clients, highly scalable server architecture, and a wide range of GIS capabilities.

ii. Enterprise Geodatabase

ArcGIS works with geographic information managed in geodatabases as well as in numerous GIS file formats. The geodatabase is the native data structure for ArcGIS and is the primary data format used for editing and data management.

Geodatabases work across a range of database management system (DBMS) architectures and file systems, come in many sizes and have varying numbers of users. The single-user databases built on files up to larger workgroup, department, and enterprise geodatabases accessed by many users.

iii. Collector for ArcGIS (Mobile Application)

Collector for ArcGIS introduces a new way to use iPhone or Android smartphone to map information. With it, the user captures and updates both tabular and spatial information using the built-in GPS capabilities of the device or by tapping on the map. Field crews use Collector to plan routes, get directions to work locations, use data-driven forms to improve data input quality, capture photos and videos of assets, and seamlessly integrate information back into the organization's GIS.

2.10 GIS DATABASE AND GIS MAPPING

To develop GPTs inventory database, a list of GPTs installed in the study will be obtained from Pejabat Lembangan Sg Klang (PLSK). Ultimately these activities provided GPTs inventory and included in gross pollutant management strategies database. This database used as a reference aid for the management of GPTs in the study area. The database can be used by the client during data collection using ArcGIS Collector. The data will be uploaded into the web base ArcGIS Online System, enable better management of gross pollutant traps, in terms of inspection & maintenance. It also allows simple analysis to help the user to see the trend of gross pollutant trapped by trapping devices.

Figure 2.7 ArcGIS Collector and ArcGIS Online By the Consultant

This application development will also include:

i. **Development of Advanced Analysis**

ii. **Development of Reporting**

iii. **Development of GPT Maintenance Module**

CHAPTER III:

GROSS POLLUTANT MANAGEMENT CASE STUDIES

3.1 INTRODUCTION

To evaluate the performance of GPTs system in this study, the results obtained from the data collection process was analysed. The analysis was done based on study objectives explained in Chapter 1. Among the results presented and discussed in this chapter includes gross pollutant wet load, gross pollutant load rates, statistical analysis, an optimum number of each GPT in each catchment, life cycle cost analysis and cost-effectiveness ratio.

3.2 GROSS POLLUTANT WET LOAD

Gross pollutants wet load were obtained from GPTs maintenance data from PLSK. The maintenance of GPTs is based on approval by PLSK, according to GPTs inspection result done by the contractor. The maintenance frequency varies from once a month or once in two months. The gross pollutant wet load data were collected and analyzed based on maintenance data from January 2015 to December 2016 for a total number of 375 proprietary GPTs that being maintained by PLSK for the whole package of ROL Project. This subchapter will discuss the cumulative wet load of each catchment supported by the descriptive analysis using SPSS. Rainfall analysis will be included in each catchment to show the relationship between rainfall and wet load. All wet load graphs of each catchment in ROL Project were developed and attached in Appendix A. Monthly rainfall versus monthly wet load graphs for each GPT is shown in Appendix B and descriptive analysis for each catchment can be referred in Appendix C.

3.2.1 Sg Klang

Figure 4.1 shows the cumulative wet load for Sungai Klang from January 2015 to December 2016. The gross pollutant wet load data were collected and analyzed based on maintenance data from January 2015 to December 2016 for a total number of 88 GPTs for Sg. Klang.

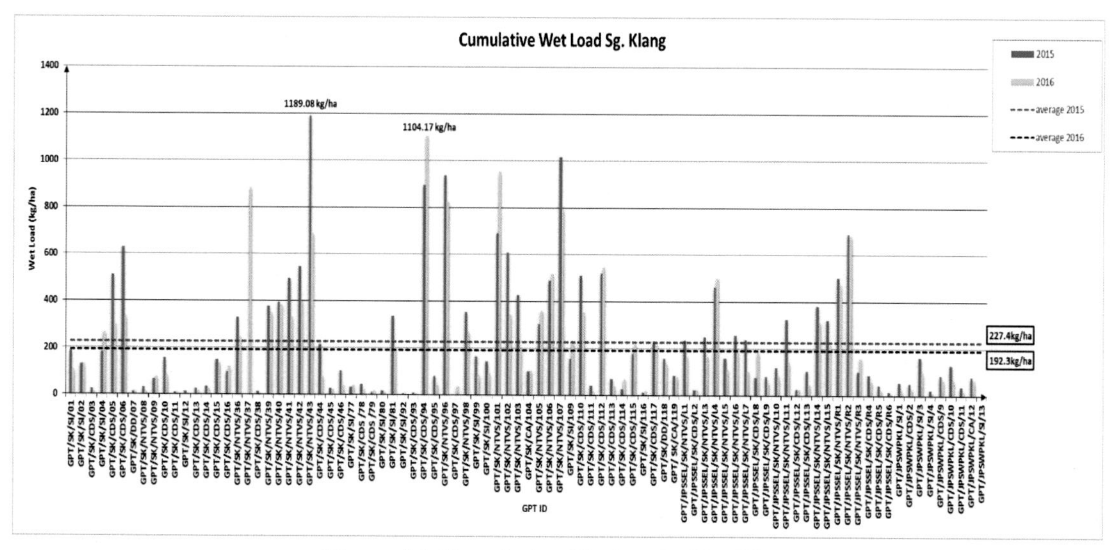

Figure 3.1 Cumulative Wet Load of Sg. Klang

Average wet load for the year 2015 and 2016 are 227.4 kg/ha and 192.3 kg/ha respectively. Average reduction for the year 2015 to 2016 is 15%. The highest of wet load are recorded at GPT/SK/NTVS/43 which is 1189.08 kg/ha for 2015 and 1104.17 kg/ha for 2016 at GPT/SS/CDS 94. Descriptive analysis for Sg Klang is discussed in Table 3.1 to show the relationship between the year 2015 and year 2016.

Table 3.1 Descriptive Analysis for Sg Klang

	N	Median	Minimum	Maximum	Interquartile Range
Wet load 2015	93	129.5100	0	1189.10	299.27
Wet load 2016	93	86.2000	0	1104.30	231.47

For year 2015, the wet load ranged from 0.00 to 1189.10 kg/ha (Mdn=129.510, IQR=299.27). Netload was non- normally distributed, with positive significantly skewed of 1.428 (SE = 0.250) and kurtosis of 2.848 (SE = .495). For year 2016, the wet load ranged from 0.00 to 1104.30 kg/ha (Mdn=86.20, IQR=231.47). Netload was non-normally distributed, with skewness of 1.975 (SE = 0.913) and kurtosis of 3.789 (SE =0.495). According to a non-parametric test, this study employed Wilcoxon signed-rank test showed that the difference between wet load years for both years is significant ($p<0.05$).

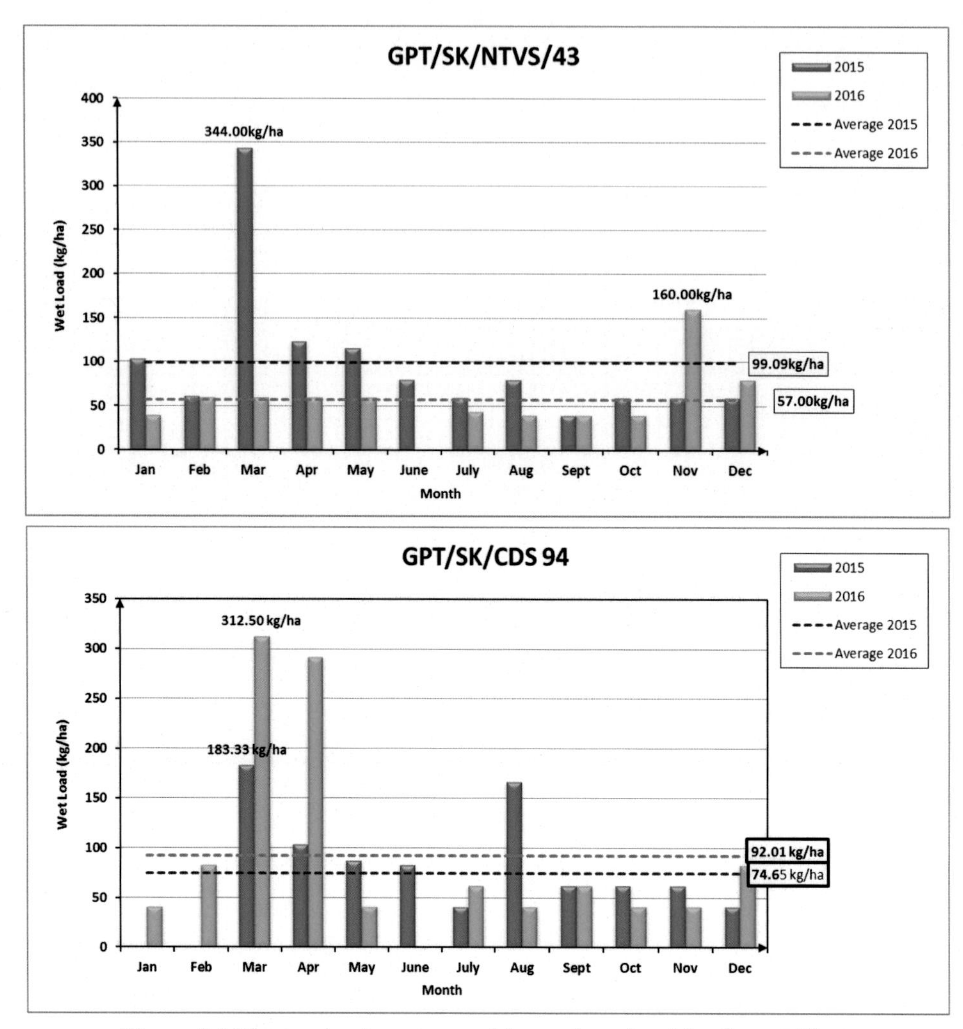

Figure 3.2 The comparison on monthly total wet load for Sungai Klang from January 2015 to December 2016 for selected GPTs

GPT/SK/NTVS/94 was installed in Sungai Klang catchment. As there are no maintenance activities in January and February 2015, the total amount of trapped gross pollutant load in March is higher compared to another month by 344 kg/ha. But, once the cleaning has been performed once a month afterward, the total wet load decreased to around 124 kg/ha. The high result wet load of GPT/SK/NTVS/43 is caused by its location within a commercial area which is near SK Ulu Kelang.

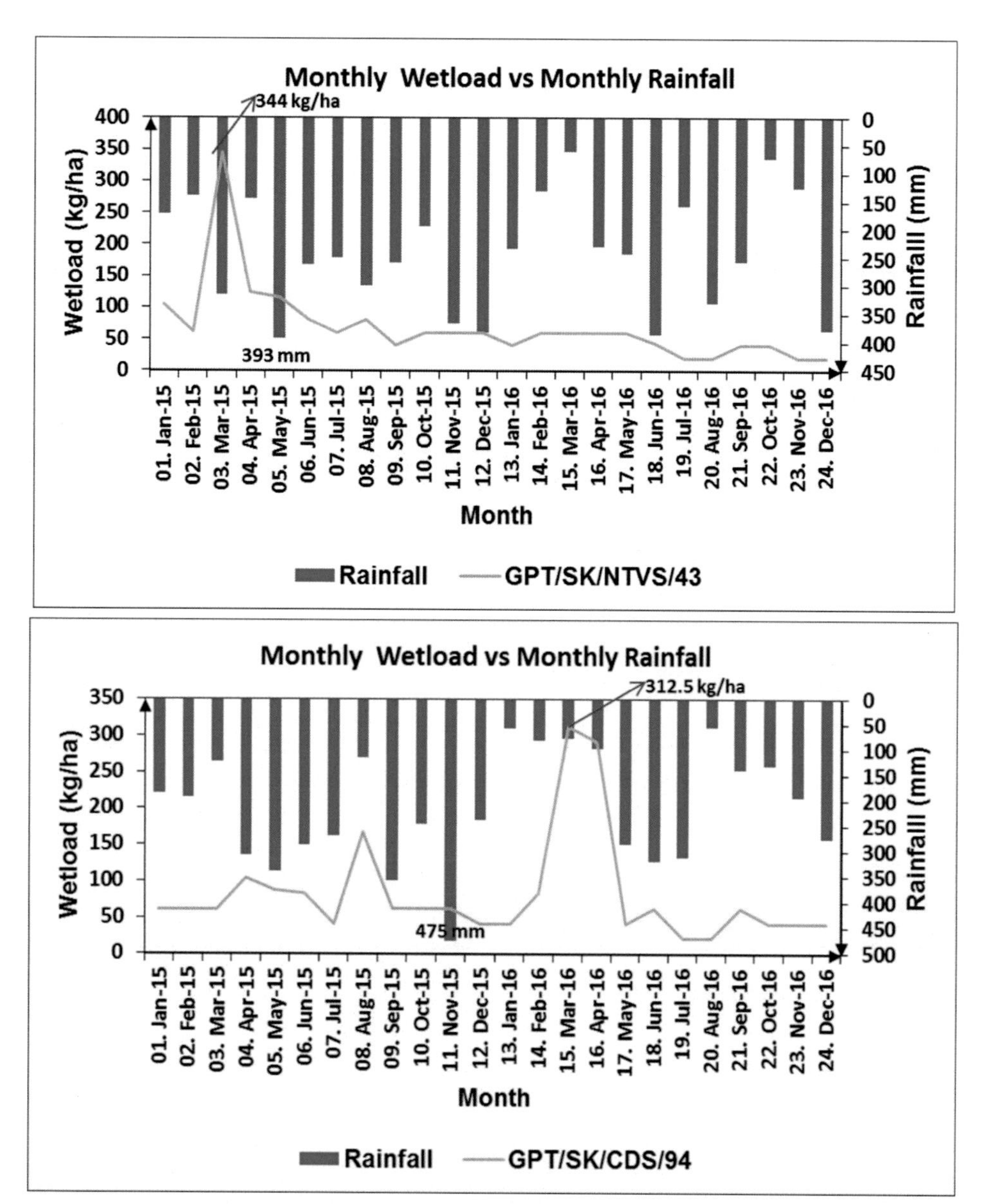

Figure 3.3 Relationship of monthly total wet load and rainfall depth for Sg Klang from January 2015 to December 2016 for selected GPTs

Figure 3.3 shows the graph relationship of monthly total wet load and rainfall depth for Sg Klang from January 2015 to December 2016 for GPT/SK/NTVS/43 and GPT/SK/CDS/94. The total amount of trapped gross pollutant load for GPT/SK/NTVS/43 in March 2015 is higher compared to another month by 344 kg/ha and the rainfall depth for that particular month is 314.5 mm. The maximum rainfall depth is in May 2015 with the reading of rainfall is 393 mm with the reading of wet load 116 kg/ha. Although the rainfall trend in GPT/SK/NTVS/43 catchment area is a variance for that entire 2 years, most of the wet load amounts are below 100 kg/ha. The total amount of trapped gross pollutant load for GPT/SK/CDS/94 in March 2016 is higher compared to another month by 312.5 kg/ha and the rainfall depth for that particular month is only 77 mm. The maximum rainfall depth is in November 2015 with the reading of rainfall is 475 mm with the reading of wet load 62.5 kg/ha.

3.2.2 *Sg Sering*

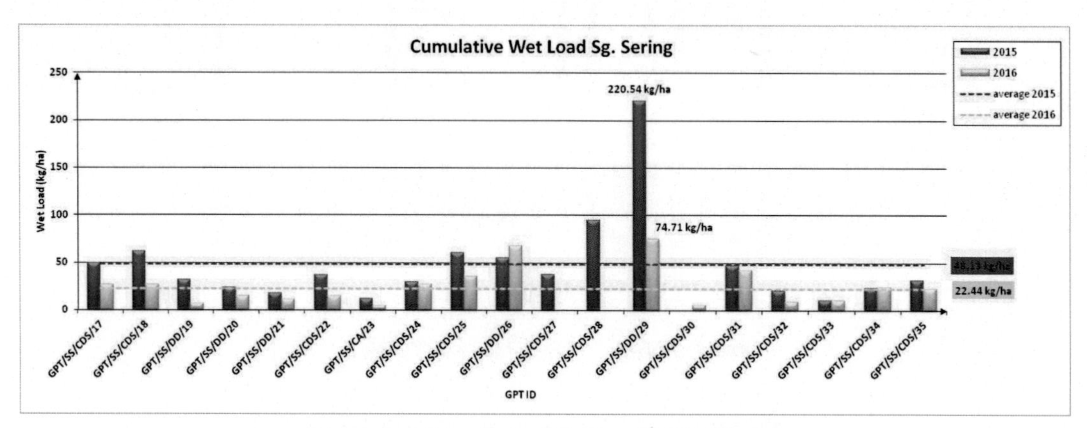

Figure 3.4 Cumulative Wet Load of Sg. Sering

Figure 3.4 shows the cumulative wet load for Sg Sering from January 2015 to December 2016. The gross pollutant wet load data were collected and analyzed based on maintenance data from January 2015 to December 2016 for a total number of 19 GPTs for Sg Sering. Average wet load for the year 2015 and 2016 are 48.13 kg/ha and 22.44 kg/ha respectively. Average reduction for the year 2015 to 2016 is 24.3%. The highest of the wet load is recorded at GPT/SS/DD/29 which is 220.54 kg/ha and 74.71 kg/ha for 2015 and 2016 respectively. Descriptive analysis for Sg Sering is discussed in Table 3.2 to show the relationship between the year 2015 and year 2016.

Table 3.2 Descriptive Analysis for Sg Sering

	N	Median	Minimum	Maximum	Interquartile Range
Wet load 2015	19	31.65	7.59	220.50	34.29
Wet load 2016	19	15.48	0.00	74.75	20.70

For the year 2015, the wet load ranged from 7.59 to 220.50 kg/ha (Mdn=31.65, IQR=34.29). The wet load was non-normally distributed, with positive significantly skewed of 3.04 (SE = 0.53) and kurtosis of 10.828 (SE = 1.014). For year 2016, the wet load ranged from 0.00 to 74.75 kg/ha (Mdn=15.48, IQR=20.70). The wet load was non-normally distributed, with skewness of 1.319 (SE = 0.524) and kurtosis of 1.658 (SE =1.014). According to a non-parametric test, this study employed Wilcoxon signed-rank test showed that the difference between wet load years for both years is significant ($p<0.05$).

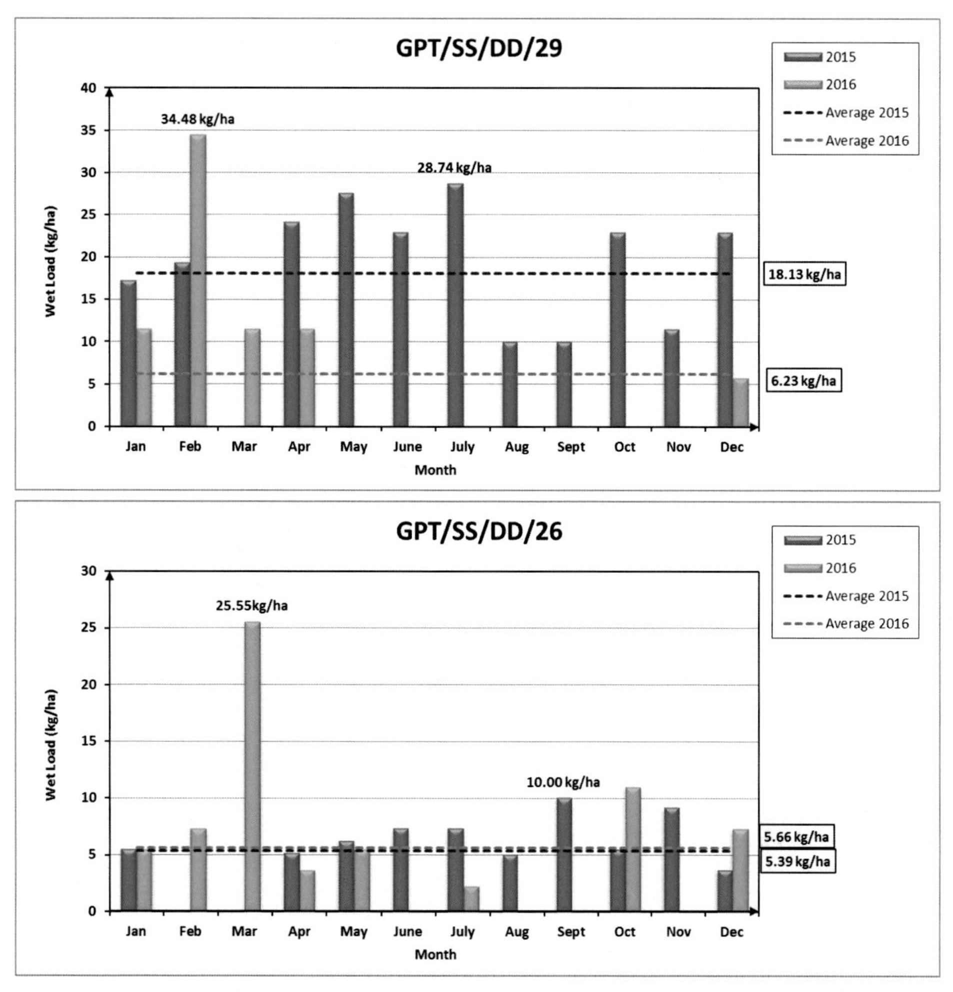

Figure 3.5 The comparison on monthly total wet load for Sg Sering from January 2015 to December 2016 for selected GPTs

GPT/SS/DD/29 was installed in Sungai Sering catchment. Average wet load for the year 2015 and 2016 are 18.13 kg/ha and 6.23 kg/ha respectively as shown in Figure 3.5. The total amount of wet load in July is highest by 28.74 kg/ha for 2015 and 34.84 kg/ha for 2016. Similar result show for 2016, the amount of wet load decrease to 11.49 kg/ha. In another hand, from May 2016 to November 2016, there is no data can be obtained. GPT/SS/DD/29 was located within the rural area and near Entrance to Taman Bukit Mulia. Average wet load GPT/SS/DD/26 for the year 2015 and 2016 are 5.39 kg/ha and 5.66 kg/ha respectively (Figure 3.5). GPT/SS/DD/26 shows the highest value 10.00 kg/ha (September 2015) and 25.55 kg/ha (March 2016).As the cleaning has been performed once a month afterward, the total wet load decreased drastically below the average value. GPT/SS/DD/26 was located within mix area and near Taman Kelab Ukay Pump House. Figure 4.9 shows the graph relationship of monthly total wet load and rainfall depth for Sg Sering from January 2015 to December 2016 for GPT/SS/DD/26 and GPT/SS/DD/29.

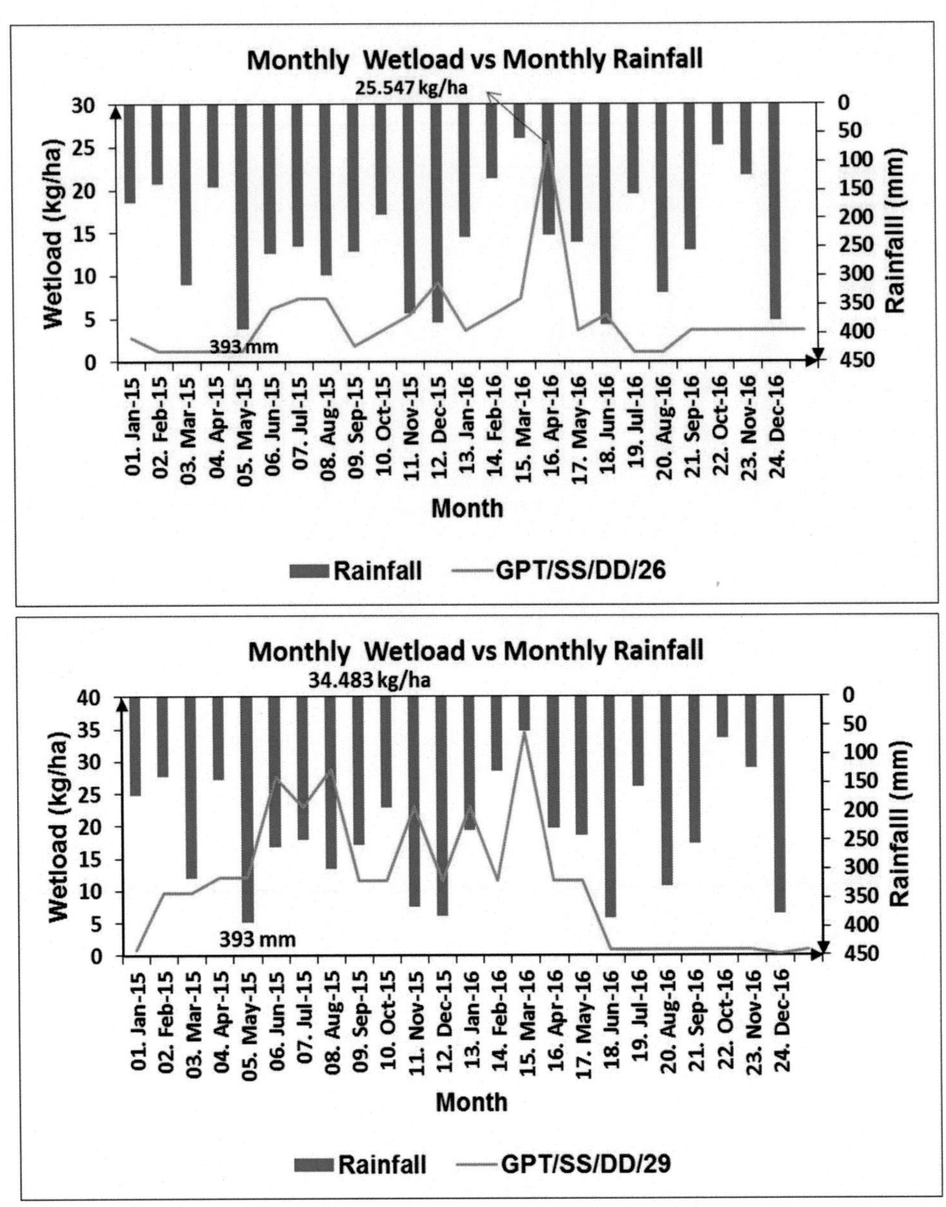

Figure 3.6 Relationship of monthly total wet load and rainfall depth for Sg Sering from January 2015 to December 2016 for selected GPTs

The total amount of trapped gross pollutant load for GPT/SS/DD/26 in May 2016 is higher compared to another month by 25.5 kg/ha and the rainfall depth for that particular month is 229.5 mm. The maximum rainfall depth is in May 2015 with the reading of rainfall is 393 mm with the reading of wet load 1.3 kg/ha. The total amount of trapped gross pollutant load for GPT/SS/DD/29 in March 2016 is higher compared to another month by 34.5 kg/ha and the rainfall depth for that particular month is only 59 mm. The maximum rainfall depth is in May 2015 with the reading of rainfall is 393 mm with the reading of wet load only 12.1 kg/ha.

3.3 SITE VERIFICATION OF SELECTED GPTS

The field visits played an integral role in this project as they were used to ensure locations of GPTs were mapped accurately as in GPTs inventory database. This process helps to a better management of GPTs inventory in the future. The objective of field validation is to ensure accurate location maps of GPTs.

In order to get accurate results and analysis on a wet load of gross pollutants after the maintenance of GPTs, the consultant has conducted a site verification to verify the actual catchment area of selected GPTs and 20 damaged GPTs to confirm the status as listed in methodology. Observations at surrounding areas, for example, the road accessibility for GPTs maintenance, the inflow and outflow of drainage of GPTs and their environmental aspects which contribute to the amount of gross pollutant trapped by the GPTs were also made during the site verification. Site verification was done on basis of site observation of each drainage line covered by each GPT for each particular area. The catchment is defined based upon a number of factors including distance to the facility, actual drainage line to the GPT, geographic boundaries within the catchment.

It is very important to ensure the catchment area is correct for every GPT design, as improper consideration of catchment area effect over design or under the design of GPTs. The site visit verification and details of each selected GPTs are attached in Appendix D.

3.4 ESTIMATION ON THE REQUIRED ADDITIONAL NUMBER OF GPTS

To estimate the number of GPTs required to be installed within RoL catchments, the values from literature (IE Aust, 2006) has been used as references for gross pollutant load in kg/ha/yr/GPT.

Based on the values from the Spring, South Africa, Cape Town, South Africa and Melbourne, the optimal annual gross pollutant wet load was calculated for each land use for all catchments. The ratio of the actual wet load by land-use to optimal annual wet load is determined to identify the numbers of additional GPTs needed.

Table 3.3 Summary Total Number of GPTs Required for Each Sub Catchments

Catchment	**Existing Total No. of GPTs**	**Additional Number of GPTs Required**		
		(Spring, South Africa, 1998)	**(Cape Town, South Africa, 2004)**	**Australia**
Sg Klang	88	93	338	197
Sg Sering	19	-	-	-

Spring, South Africa: a highly industrial city
Cape Town, South Africa: highly urbanized city
Australia: highly urbanized city

3.5 LCC OF GPT

The study area for LCC is limited to Sg Klang Upper and Sg Kerayong only for this report due to lack of data for another catchment. The LCC for all GPT in the study area has been analyzed. The analysis includes EAC derived from the LCC, based on different project duration and divided into 3 types of maintenance method which are suction truck, using manual and using a crane. The project duration selected was 1, 10 and 20 years based on the literature review.

From the analysis, LCC of GPTs for the suction truck method in the study area is ranging from RM 113,130 to RM 270,686 for the duration of 10 years. However, for project duration of 20 years, the LCC of GPTs ranged from RM RM 138,092.00 to RM 350,822.00

For the manual method, LCC of GPTs is ranging from RM 122,726.00 to RM 157, 750.00 for duration 10 years and starting from RM 140,238.00 up to RM 210,286.00 for duration 20 years. While for the crane method, LCC of GPTs for the suction truck method in the study area is ranging from RM 226,062.00 to RM 261,574.00 for the duration of 10 years. However, for project duration of 20 years, the LCC of GPTs ranged from RM 297,086.00 to RM 403,622.00.

Figure 3.7 until Figure 3.9 shows the percentage of GPTs with a different range of LCC based on their maintenance method. The detailed analysis of LCC is shown in Appendix E.

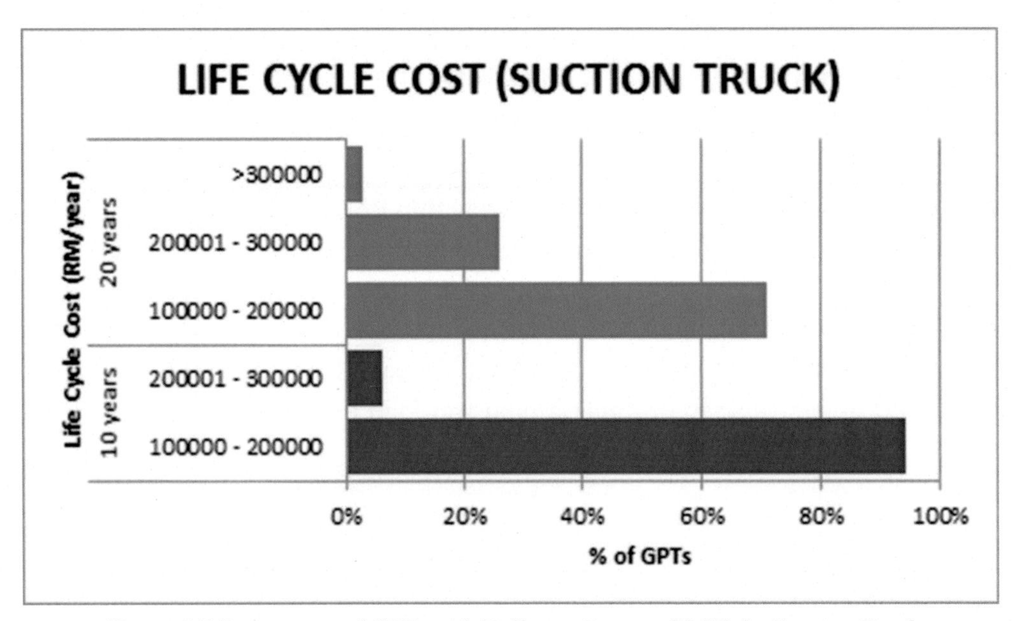

Figure 3.7 Percentage of GPTs with Different Range of LCC for Suction Truck

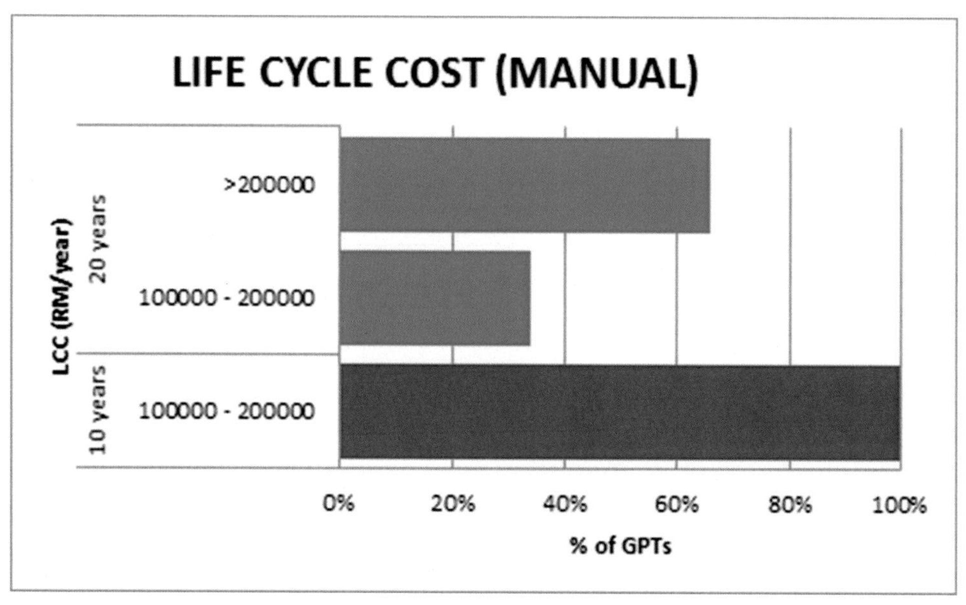

Figure 3.8 Percentage of GPTs with Different Range of LCC for Manual

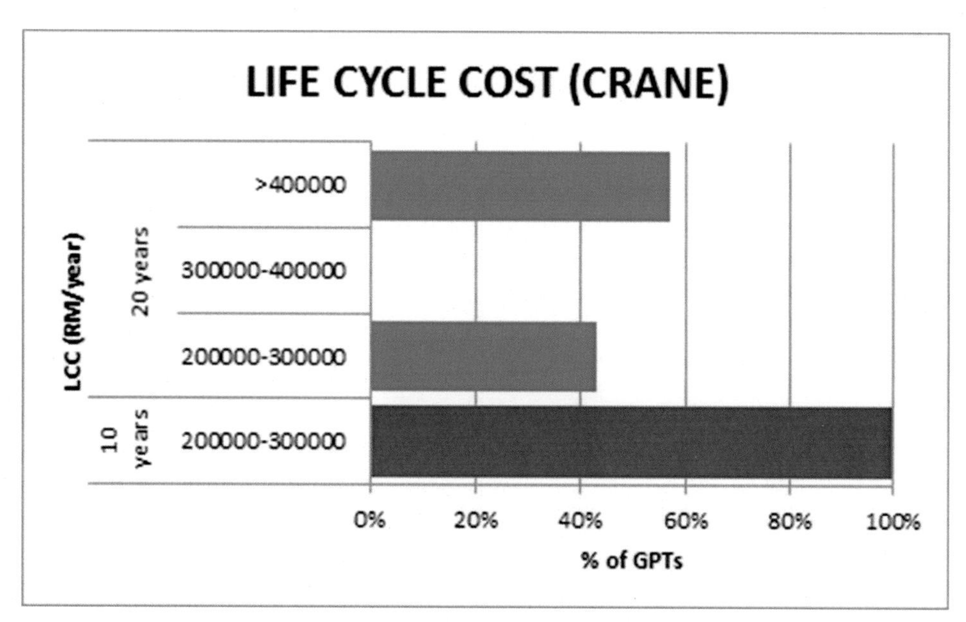

Figure 3.9 Percentage of GPTs with Different Range of LCC for Crane

3.6 CER OF GPT

Another important criterion in assessing the performance of GPTs is analysed in terms of cost. The analysis involved is quantifying the value for money for each device, using the "Cost-Effectiveness Ratio" analytical technique (Brisbane City Council 2002). The CER takes into account the LCC and PRE. The important advantage of the CER is that it provides a simple tool to assess management options for pollution trap operations. The CER for all 151 GPTs in the study area was calculated, and the detail result is shown in Appendix E.

Figure 3.10 shows the top CER for each GPT based on their maintenance method. Normally, there is 5 type of GPTs are using maintenance method of the suction truck which are CDS, SI, DD, HUME, PVT and ECO. However, on average, PVT is having the lowest CER with a value of RM 30.74 per kg per year for the suction truck method. While only NTVS is using manual method and NTVS is having CER value of RM 36.65. For crane method, CA is quite expensive with CER value of RM 369.72. Generally, devices with the lowest CER are preferred, meaning that they captured more pollutants and were not excessively priced to clean (Dean, 2007).

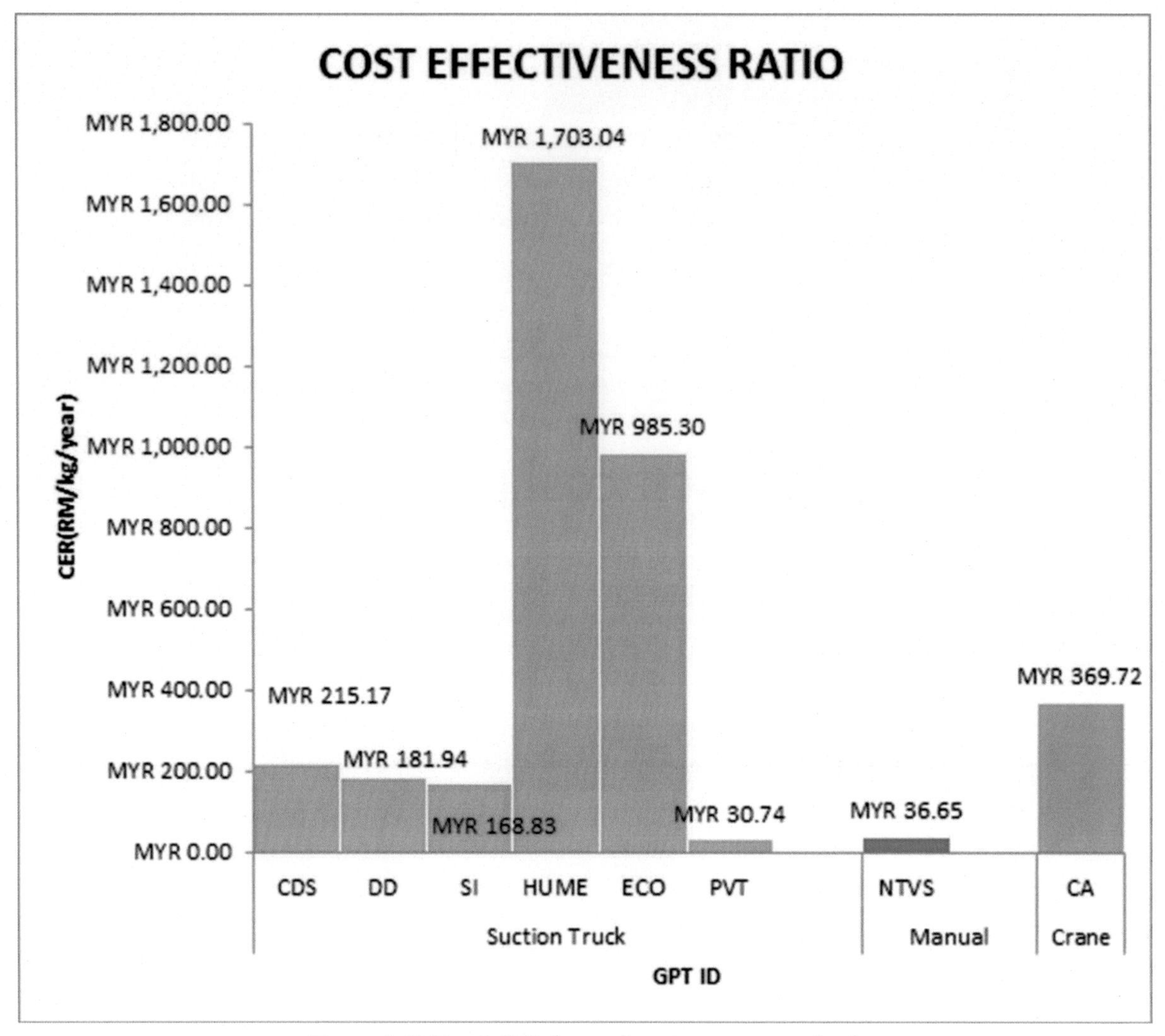

Figure 3.10 Top 3 CER for each maintenance method.

CHAPTER IV:

CONCLUSION AND RECOMMENDATIONS

4.1 CONCLUSION

The aim of this study is to improve the understanding of the load of gross pollutant trapped in ROL catchment area. Apart from that, simple calculation on the optimum number is calculated by comparing with other countries which are Spring and Cape Town, South Africa and Australia. In order to optimize the maintenance cost and cost-effectiveness of GPTs, LCC and CER are calculated. Finally, all the information will be stored in gross pollutant management strategies knowledge database system to ensure that it is effective and efficient for DID to find and use the necessary data.

Ultimately, the data collected assist the engineers and local authorities to implement appropriate strategies for trapping gross pollutants in an urban area, expand the sources for managing gross pollutants in order to rehabilitate the river system and preparing budget allocation of using GPTs for future operation & maintenance. The conclusions of this study are as follows:

- The study review and enhance the existing GPTs inventory system, planning & management tool for managing GPTs. Based on the analysis, only Sg Bunus and Sg Keroh catchment area proved that the amount of wet load are increased when the rainfall depth is increased while the other catchment area shows no significant trends. The conclusion can be made is there are no significant trends of increasing gross pollutant wet load to the stormwater conveyance system with increasing rainfall depth. The gross pollutant wet loads for this study area are not influenced by the rainfall depth as the main factor, other aspects such as behavior of the resident in that particular catchment area, the land use characteristic and the sizing of GPTs type need to be considered as the influences factor. Another factor is localized rain event which often happens in Malaysia. From the analysis, Sg Sering ($y = 1.0115x + 27.939$, $R2=0.975$) shows a very strong relationship between total wet load and rainfall. From this study, it can be concluding that the value of wet load will not be affected by rainfall intensity. During site verification, it was found that it is essential to have the right catchment design in order to ensure optimum performance of GPT as well as minimizing LCC. All catchments required additional GPTs except for Sg Sering.
- The study optimize the maintenance cost in effective ways without affecting the normal maintenance operation system and structure. From the analysis, LCC of GPTs for the suction truck method in

the study area is ranging from RM 113,130 to RM 270,686 for 10 years. However, for a project duration of 20 years, the LCC of GPTs ranged from RM 138,092.00 to RM 350,822.00. For the manual method, LCC of GPTs is ranging from RM 122,726.00 to RM 157, 750.00 for duration ten years and starting from RM 140,238.00 up to RM 210,286.00 for duration 20 years. While for the crane method, LCC of GPTs for the suction truck method in the study area is ranging from RM 226,062.00 to RM 261,574.00 for 10 years. However, for a project duration of 20 years, the LCC of GPTs ranged from RM 297,086.00 to RM 403,622.00. Besides, for CER analysis, PVT is having the lowest CER with a value of RM 30.74 per kg per year for the suction truck method. While only NTVS is using manual method and NTVS is having CER value of RM 36.65. For crane method, CA is quite expensive with CER value of RM 369.72. Frequency maintenance is developed based on 2 years wet load with range:

- 0 kg to 10 kg – monthly,
- 11 kg to 29 kg – Every 2 month
- 30 kg above – Every 3 month

4.2 FUTURE RECOMMENDATIONS

To ensure the effective management of gross pollutants, the following suggestions are recommended:

- To collect more local data such as the volume of gross pollutants, to enable a more accurate estimation of GPTs maintenance.
- To conduct a more comprehensive study on the optimum number of GPTs as a measure to reduce the amount of rubbish traveled into the river system.
- Implementation of the non-structural method (as recommended by MSMA), through public awareness regarding the importance of preserving nature and avoiding pollutants, shall be actively done by all parties involved to reduce the amount of debris produced by year.
- Local authorities should be more proactive and implement the necessary acts and regulations to sustain the quality of the environment.
- Education provided through the medium of mass media, seminars, courses, and any other ways to the young generation to preserve the nature and environment.

REFERENCES

(ROL-POP), R. o. (2012 - 2015). *River of Life.* Retrieved from River of Life Website:http://www.myrol.my/index.cfm?&menuid=96

A. Ab Ghani, N. Armitage, S. Pithey, *A Study of the litter loadings in urban drainage systems – methodology and objectives*, Water Sci. technol., 44 (2001) 1117-1127.

Allison, R., Chiew, F. and Mcmahon, T. (1997a). *Stormwater Gross Pollutants Industry Report 97/11*, Cooperative Research Centre for Catchment Hydrology, Monash University, Australia.

Allison, R.A. and Chiew, F.H.S., (1995). *Monitoring Of Stormwater Pollution for Various Land-Uses in an Urban Catchment*, Proceeding 2nd International Symposium On Urban Stormwater Management, Melbourne Australia IE Aust., NCP 95/03, Vol.2, p. 511-516.

Allison, R.A., Chiew, F. and Mcmahon, T. (1998). *A Decision Support System For Determining Effective Trapping Strategies For Gross Pollutants*, Technical Report 98/3, Cooperative Research Centre For Catchment Hydrology, Monash University, Australia.

Allison, R.A., Rooney, G.R., Chiew, F.H.S. And Mcmahon, T.A. (1997b). *Field Trials of Side Entry Pit Traps for Urban Stormwater Pollution Control*, Proceedings of the 9th National Local Government Engineering Conference, Institution of Engineers Australia, Melbourne, Australia.

Armitage, N. (2006). *The Removal of Urban Solid Waste from Stormwater Drains,* Department of Civil Engineering, University of Cape Town, South Africa

Armitage, N.P., Rooseboon, A., Nel, C. and Townshend, P., (1998). *The Removal of Urban Litter from Stormwater Conduits and Streams*, Report To Water Research Commission Of South Africa, Report No. 95/98.

Armitage, N. 2003. The removal of urban solid waste from stormwater drains. Prepared for the International Workshop on Global Developments in Urban Drainage Management, Indian Institute of Technology, Bombay, Mumbai India. 5-7 February.

Armitage, N. 2007. The reduction of urban litter in the stormwater drains of South Africa. Urban Water Journal Vol. 4, No. 3: 151-172. September.

Asian Development Bank (2007), *Malaysia: Klang River Basin Environment Improvement and Flood Mitigation Project.*

AS/NZS 4536: 1999 *Life Cycle Costing – An Application Guide*, Standard s Australia.

Brisbane City Council (2002). *'SQUID Monitoring Program Stage 4 2000/2001',* Technical Report Revision 1 by Water and Environment City Design

Caltrans (2002). *Continuous Deflective Separation Units - Operations, Maintenance & Monitoring CTSW-RT-02-037,* California Department of Transportation Sacramento.

Cascadia Consulting Group, Inc. 2005. Washington State Litter Study – Litter Generation & Composition Report. Washington State Department of Ecology. March.

Collins, A. (2003). *Catchment Remediation Capital Works Annual Performance Report*, Hornsby Shire Council, New South Wales, Australia.

Cornelius, M., Clayton, T., Lewis, G., Arnold, G. And Craig, J. (1994). *Litter Associated With Stormwater Discharge in Auckland City New Zealand*, Auckland.

CSIRO (2006). *Chapter 7 – Structural Treatment Measures*, Urban Stormwater: Best Practice Environmental Management Guidelines.

Dean, D. K., Natalie, P., and Murray, P., (2007). *Auditing and Maintenance Costing of Gross Pollutant Traps, Protecting our Urban Waterways - Water Quality and Managing Flow Impacts,* Blacktown City Council, Australia.

Department Of Environment and Conservation New South Wales (2004). *Evaluation Of The NSW Urban Stormwater Education Program – The Drain Is Just For Rain*, Summary Report, New South Wales.

Department of Water (2006). Government of Western Australia, Website: *www.**water**.wa.gov.au/PublicationStore/first/84993.pdf*, reviewed on December 2015

DID. (2012), Chapter 10, *Storm Water Management Manual For Malaysia 2nd Edition*, Department of Irrigation and Drainage (DID), Malaysia.

Dodge, C. (2005). *AMAFCA/Albuquerque MS4 Floatable and Gross Pollutant Study*, Engineers Australia 2006, *Australian Runoff Quality – A Guide to Water Sensitive Urban Design*, Engineers Media, Crow's Nest, New South Wales.

England, G. and Rushton, B. (2003). *An Update on ASCE Monitoring Guidelines for Measuring Stormwater Gross Solids*, Florida.

Federal Highway Administration (FHWA) (2002). Stormwater Best Management Practices In An Ultra-Urban Setting: Selection And Monitoring, U.S Department Of Transportation.

Fitzgerald, B. and Bird, W. (2010). *Literature Review: Gross Pollutant Traps as a Stormwater Management Practice*. Auckland Council Technical Report 2011/006.

Goonetilleke, A. and Thomas, E. C. (2003). *Urban Water Quality And The Triple Bottom Line - Can We Reconcile The Irreconcilables?*, In Search of Sustainability.

Hornsby Shire Council (2003). *The Catchment Remediation Capital Works Program*, Annual Performance Report 2002/2003, Hornsby NSW 1630.

Island Care New Zealand Trust (1996). *Reducing the Incidence of Stormwater Debris and Street Litter in the Marine Environment - A Co-Operative Community Approach,* Auckland.

J.T. Madhani, J. Young R.J. Brown, *Visualizing experimental flow fileds through a stormwater gross pollutant trap*, J. Visualiztion, 17 (2014) 17-26.

Kim, L.H, M. Kayhanian, M.K. Stenstrom 2004. Event mean concentration and loading of litter from highways during storms. Science of the Total Environment Vol 330: 101-113.

Laist, D. W. and M. Liffmann 2000. Impacts of marine debris: research and management needs. Issue papers of the International Marine Debris Conference. pp 16-29. Honolulu, HI. 6-11 Aug.

Land Development Guidelines. (2007). *Our Living City. Gold Coast: Gold Coast Planning Scheme Policies.*

Lariyah, M. S., Mohd Nor, M.D., Mohamad Khairudin, K., Chua, K.H., Norazli, O., and Leong, W. K. (2006). Development of Stormwater Gross Pollutant Traps (GPT's) Decision Support System For River Rehabilitation, National Conference – Water For Sustainable Development Towards A Developed Nation By 2020, Guoman Resort, Port Dickson.

Lariyah, M.S., Mohd Nor M.D., Norazli O., Md. Nasir M.N., Hidayah B., and Zuleika Z. (2011). *Gross Pollutants Analysis in Urban Residential Area for a Tropical Climate Country,* 12th International Conference on Urban Drainage, Porto Alegre/Brazil.

Lariyah, M.S., Takara, K., Aminuddin A.G., Azazi, Z., Rozi, A. and Mohd Nor, M.D. (2004). *A Life Cycle Cost (LCC) Assessment of Sustainable Urban Drainage System Facilities*, 1st International Conference on Managing Rivers in the 21st Century: Issues and Challenge – Rivers'04, Malaysia.

Lewis, J. 2002. Effectiveness of stormwater litter traps for syringe and litter removal. Cooperative Research Centre for Catchment Hydrology, Melbourne.

Lippner, G., R. Churchwell, R. Allison, G. Moeller, and J. Johnston 2001. A Scientific Approach to Evaluating Storm Water Best Management Practices for Litter. Transportation Research Record . TTR 1743, 10-15.

Los Angeles County 2002. Los Angeles County Litter Monitoring Plan for the Los Angeles River and Ballona Creek Trash Total Maximum Daily Load. 30 May.

Los Angeles County 2004a. Trash Baseline Monitoring Results Los Angeles River and Ballona Creek Watershed. Los Angeles County Department of Public Works. 17 February.

Los Angeles County 2004b. Trash Baseline Monitoring for Los Angeles River and Ballona

Creek Watersheds. Los Angeles County Department of Public Works. 6 May.

Los Angeles County 2005. Los Angeles County 1994-2005 Integrated Receiving Water Impacts Report, Final Report - August 2005. Available at http://dpw.lacounty.gov/wmd/NPDES/1994-05_report/contents.html. Accessed Dec 6, 2010.

Los Angeles Regional Board (California Regional Water Quality Control Board – Los Angeles Region) 2007. Trash Total Maximum Daily Loads for the Los Angeles River Watershed. Revised Draft. 27 July.

M.A. Wilson, O. Mohseni, J.S. Gulliver, R.M. Hozalki, H.G. Stefan, *Assesment of hydrodynamics separators for storm-water treatment*, J. Hydraul. Eng., 135 (2009) 383-392.

Marais, M., Armitage, N. and Pithey, S. (2001). *A Study of the Litter Loadings in Urban Drainage Systems – Methodology and Objectives*, Water Science Technology Vol. 44 No. 6, IWA Publishing, p. 99-108.

NSW EPA (2000) New South Wales Environmental Protection Authority (2000). *Unpublished Costing Data For Structural Stormwater Quality BMPs Derived from NSW Local Governments*, NSW EPA, Sydney, New South Wales.

Q. Zhang, J.L. Zhao, G.G. Ying, Y.S. Liu, C.G. Pan, *Emission estimation and multimedia fate modelling of seven steroids at the river scale basin scale in China*, Environ. Sci. Technol., 48 (2014) 7892-7992.

RBF Consulting (2003). *Trash and Debris – Best Management Practice (BMP) Evaluation*, Appendix E2, Orange County Stormwater Program, Irvine, CA9261.

Roesner, L.A., Pruden, A. and Kidner, E.M (2007). *Improved Protocol For Classification And Analysis Of Stormwater Borne Solid*, Water Environment Research Foundation, Colorado State University.

Rushton, B., England, G. and Smith, D. (2007). *ASCE Guideline for Monitoring Stormwater Gross Pollutants,* Draft.

Sarizah, L. (2006). *Kajian Terhadap Penggunaan Perangkap Sampah Dalam Sistem Saliran Terbuka*, Bachelor of Civil Engineering Thesis, Universiti Teknologi Malaysia.

Siti Norazela, H. (2007). *Study in Assessing Performance of Gross Pollutant Traps in Open Channel*, Bachelor of Civil Engineering Thesis, Universiti Teknologi Malaysia.

T.H.F. Wong, W. Tracey, *Peer Review and Development of a Stormwater Gross Pollutant Treatment Technology Assessment Methodology*, Monash University, Australia, 2002.

Taylor, A.C. (2003). *An Introduction to Life Cycle Costing Involving Structural Stormwater Quality Management Measures*, Cooperative Research Centre For Catchment Hydrology, Australia.

Taylor, A.C. (2005). *Structural Stormwater Quality BMP Cost / Size Relationship Information from the Literature*, Cooperative Research Centre for Catchment Hydrology, Australia.

Printed in the United States
By Bookmasters